Table of contents

Section 1

Name:...................................   Date:...................................

Find the sum

Complete all the activities (Addition).

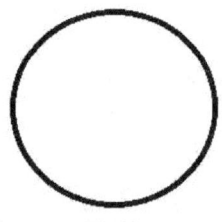

SCORE

1. 600 + 191	2. 981 + 644	3. 700 + 938	4. 207 + 900	5. 372 + 367	6. 959 + 819	7. 489 + 460
8. 186 + 866	9. 856 + 542	10. 214 + 199	11. 463 + 482	12. 776 + 535	13. 816 + 147	14. 497 + 835
15. 885 + 647	16. 436 + 127	17. 168 + 153	18. 407 + 196	19. 340 + 494	20. 360 + 596	21. 630 + 372
22. 324 + 791	23. 457 + 437	24. 132 + 851	25. 478 + 851	26. 371 + 831	27. 742 + 138	28. 163 + 218
29. 414 + 546	30. 827 + 652	31. 542 + 371	32. 115 + 724	33. 930 + 932	34. 751 + 437	35. 153 + 989
36. 626 + 462	37. 282 + 616	38. 568 + 191	39. 258 + 984	40. 179 + 479	41. 115 + 645	42. 457 + 135

Name:................................ Date:................................

Find the sum

Complete all the activities (Addition).

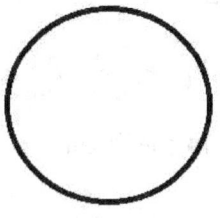

SCORE

1. 600 + 191 791	2. 981 + 644 1 625	3. 700 + 938 1 638	4. 207 + 900 1 107	5. 372 + 367 739	6. 959 + 819 1 778	7. 489 + 460 949
8. 186 + 866 1 052	9. 856 + 542 1 398	10. 214 + 199 413	11. 463 + 482 945	12. 776 + 535 1 311	13. 816 + 147 963	14. 497 + 835 1 332
15. 885 + 647 1 532	16. 436 + 127 563	17. 168 + 153 321	18. 407 + 196 603	19. 340 + 494 834	20. 360 + 596 956	21. 630 + 372 1 002
22. 324 + 791 1 115	23. 457 + 437 894	24. 132 + 851 983	25. 478 + 851 1 329	26. 371 + 831 1 202	27. 742 + 138 880	28. 163 + 218 381
29. 414 + 546 960	30. 827 + 652 1 479	31. 542 + 371 913	32. 115 + 724 839	33. 930 + 932 1 862	34. 751 + 437 1 188	35. 153 + 989 1 142
36. 626 + 462 1 088	37. 282 + 616 898	38. 568 + 191 759	39. 258 + 984 1 242	40. 179 + 479 658	41. 115 + 645 760	42. 457 + 135 592

Name:.................................   Date:.................................

Find the sum

Complete all the activities (Addition).

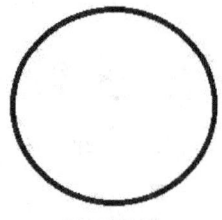

SCORE

1. 665 + 773	2. 891 + 843	3. 907 + 819	4. 941 + 865	5. 321 + 960	6. 691 + 548	7. 326 + 581
8. 759 + 249	9. 688 + 255	10. 883 + 833	11. 650 + 553	12. 600 + 808	13. 582 + 922	14. 561 + 276
15. 758 + 990	16. 530 + 944	17. 318 + 525	18. 157 + 965	19. 114 + 283	20. 195 + 534	21. 209 + 742
22. 408 + 375	23. 732 + 821	24. 185 + 781	25. 695 + 797	26. 254 + 819	27. 310 + 383	28. 816 + 212
29. 272 + 183	30. 362 + 984	31. 357 + 765	32. 207 + 330	33. 530 + 746	34. 330 + 800	35. 524 + 794
36. 235 + 505	37. 159 + 449	38. 723 + 907	39. 245 + 326	40. 176 + 760	41. 838 + 779	42. 580 + 960

Find the sum

Name:............................   Date:............................

Complete all the activities (Addition).

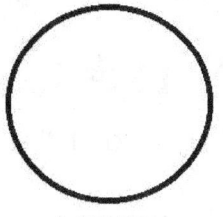

SCORE

1.	665	2.	891	3.	907	4.	941	5.	321	6.	691	7.	326
	+ 773		+ 843		+ 819		+ 865		+ 960		+ 548		+ 581
	1 438		1 734		1 726		1 806		1 281		1 239		907

8.	759	9.	688	10.	883	11.	650	12.	600	13.	582	14.	561
	+ 249		+ 255		+ 833		+ 553		+ 808		+ 922		+ 276
	1 008		943		1 716		1 203		1 408		1 504		837

15.	758	16.	530	17.	318	18.	157	19.	114	20.	195	21.	209
	+ 990		+ 944		+ 525		+ 965		+ 283		+ 534		+ 742
	1 748		1 474		843		1 122		397		729		951

22.	408	23.	732	24.	185	25.	695	26.	254	27.	310	28.	816
	+ 375		+ 821		+ 781		+ 797		+ 819		+ 383		+ 212
	783		1 553		966		1 492		1 073		693		1 028

29.	272	30.	362	31.	357	32.	207	33.	530	34.	330	35.	524
	+ 183		+ 984		+ 765		+ 330		+ 746		+ 800		+ 794
	455		1 346		1 122		537		1 276		1 130		1 318

36.	235	37.	159	38.	723	39.	245	40.	176	41.	838	42.	580
	+ 505		+ 449		+ 907		+ 326		+ 760		+ 779		+ 960
	740		608		1 630		571		936		1 617		1 540

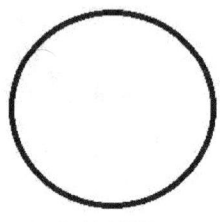

Find the sum

Complete all the activities (Addition).

1. 835 + 710	2. 734 + 934	3. 708 + 944	4. 792 + 930	5. 920 + 211	6. 276 + 443	7. 635 + 178
8. 844 + 985	9. 590 + 384	10. 357 + 900	11. 662 + 748	12. 760 + 859	13. 659 + 909	14. 966 + 946
15. 107 + 601	16. 246 + 577	17. 285 + 246	18. 494 + 819	19. 112 + 396	20. 192 + 579	21. 102 + 304
22. 584 + 948	23. 273 + 380	24. 372 + 167	25. 481 + 190	26. 269 + 554	27. 522 + 679	28. 398 + 187
29. 949 + 874	30. 375 + 798	31. 977 + 846	32. 923 + 419	33. 964 + 698	34. 258 + 489	35. 683 + 125
36. 972 + 295	37. 292 + 777	38. 898 + 458	39. 264 + 811	40. 687 + 379	41. 637 + 900	42. 565 + 916

Find the sum

Complete all the activities (Addition).

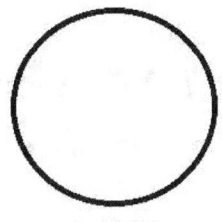

SCORE

| 1. | 835
+ 710
1 545 | 2. | 734
+ 934
1 668 | 3. | 708
+ 944
1 652 | 4. | 792
+ 930
1 722 | 5. | 920
+ 211
1 131 | 6. | 276
+ 443
719 | 7. | 635
+ 178
813 |

8. 844   9. 590   10. 357   11. 662   12. 760   13. 659   14. 966
    + 985   + 384   + 900   + 748   + 859   + 909   + 946
    1 829   974   1 257   1 410   1 619   1 568   1 912

15. 107   16. 246   17. 285   18. 494   19. 112   20. 192   21. 102
    + 601   + 577   + 246   + 819   + 396   + 579   + 304
    708   823   531   1 313   508   771   406

22. 584   23. 273   24. 372   25. 481   26. 269   27. 522   28. 398
    + 948   + 380   + 167   + 190   + 554   + 679   + 187
    1 532   653   539   671   823   1 201   585

29. 949   30. 375   31. 977   32. 923   33. 964   34. 258   35. 683
    + 874   + 798   + 846   + 419   + 698   + 489   + 125
    1 823   1 173   1 823   1 342   1 662   747   808

36. 972   37. 292   38. 898   39. 264   40. 687   41. 637   42. 565
    + 295   + 777   + 458   + 811   + 379   + 900   + 916
    1 267   1 069   1 356   1 075   1 066   1 537   1 481

Find the sum

Complete all the activities (Addition).

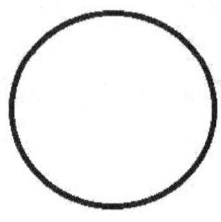

SCORE

1. 432 + 691	2. 155 + 605	3. 847 + 406	4. 868 + 238	5. 121 + 393	6. 601 + 255	7. 379 + 266
8. 983 + 574	9. 222 + 882	10. 482 + 456	11. 217 + 974	12. 362 + 744	13. 779 + 939	14. 923 + 238
15. 287 + 995	16. 261 + 177	17. 381 + 108	18. 986 + 373	19. 643 + 652	20. 641 + 377	21. 256 + 835
22. 274 + 898	23. 277 + 743	24. 179 + 997	25. 827 + 213	26. 257 + 754	27. 517 + 209	28. 799 + 457
29. 429 + 410	30. 825 + 907	31. 482 + 730	32. 950 + 125	33. 232 + 523	34. 873 + 372	35. 279 + 710
36. 929 + 568	37. 341 + 205	38. 300 + 381	39. 800 + 463	40. 567 + 136	41. 860 + 925	42. 410 + 174

Find the sum

Complete all the activities (Addition).

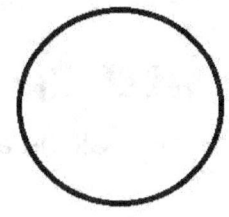

SCORE

1. 432 + 691 1 123	2. 155 + 605 760	3. 847 + 406 1 253	4. 868 + 238 1 106	5. 121 + 393 514	6. 601 + 255 856	7. 379 + 266 645
8. 983 + 574 1 557	9. 222 + 882 1 104	10. 482 + 456 938	11. 217 + 974 1 191	12. 362 + 744 1 106	13. 779 + 939 1 718	14. 923 + 238 1 161
15. 287 + 995 1 282	16. 261 + 177 438	17. 381 + 108 489	18. 986 + 373 1 359	19. 643 + 652 1 295	20. 641 + 377 1 018	21. 256 + 835 1 091
22. 274 + 898 1 172	23. 277 + 743 1 020	24. 179 + 997 1 176	25. 827 + 213 1 040	26. 257 + 754 1 011	27. 517 + 209 726	28. 799 + 457 1 256
29. 429 + 410 839	30. 825 + 907 1 732	31. 482 + 730 1 212	32. 950 + 125 1 075	33. 232 + 523 755	34. 873 + 372 1 245	35. 279 + 710 989
36. 929 + 568 1 497	37. 341 + 205 546	38. 300 + 381 681	39. 800 + 463 1 263	40. 567 + 136 703	41. 860 + 925 1 785	42. 410 + 174 584

Name:................................. Date:...............................

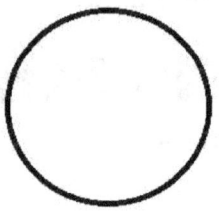

Find the sum

Complete all the activities (Addition).

1. 837 + 507	2. 659 + 272	3. 308 + 628	4. 943 + 333	5. 826 + 734	6. 894 + 524	7. 884 + 248

8. 766 + 752	9. 869 + 806	10. 818 + 218	11. 283 + 688	12. 513 + 412	13. 907 + 518	14. 656 + 966

15. 843 + 789	16. 146 + 557	17. 420 + 876	18. 287 + 724	19. 864 + 701	20. 164 + 649	21. 766 + 697

22. 741 + 606	23. 310 + 775	24. 496 + 652	25. 695 + 814	26. 266 + 154	27. 207 + 292	28. 890 + 249

29. 993 + 346	30. 364 + 789	31. 971 + 976	32. 489 + 138	33. 687 + 437	34. 204 + 108	35. 960 + 143

36. 888 + 268	37. 742 + 269	38. 451 + 585	39. 946 + 906	40. 202 + 232	41. 122 + 104	42. 370 + 129

Find the sum

Complete all the activities (Addition).

Name:................................ Date:................................

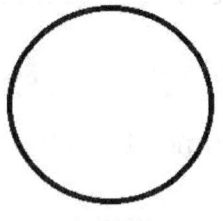

SCORE

1.	837 + 507 1 344	2.	659 + 272 931	3.	308 + 628 936	4.	943 + 333 1 276	5.	826 + 734 1 560	6.	894 + 524 1 418	7.	884 + 248 1 132

1.   837   2.   659   3.   308   4.   943   5.   826   6.   894   7.   884
   + 507   + 272   + 628   + 333   + 734   + 524   + 248
  1 344   931   936   1 276   1 560   1 418   1 132

8.   766   9.   869   10. 818   11. 283   12. 513   13. 907   14. 656
   + 752   + 806   + 218   + 688   + 412   + 518   + 966
  1 518   1 675   1 036   971   925   1 425   1 622

15. 843   16. 146   17. 420   18. 287   19. 864   20. 164   21. 766
   + 789   + 557   + 876   + 724   + 701   + 649   + 697
  1 632   703   1 296   1 011   1 565   813   1 463

22. 741   23. 310   24. 496   25. 695   26. 266   27. 207   28. 890
   + 606   + 775   + 652   + 814   + 154   + 292   + 249
  1 347   1 085   1 148   1 509   420   499   1 139

29. 993   30. 364   31. 971   32. 489   33. 687   34. 204   35. 960
   + 346   + 789   + 976   + 138   + 437   + 108   + 143
  1 339   1 153   1 947   627   1 124   312   1 103

36. 888   37. 742   38. 451   39. 946   40. 202   41. 122   42. 370
   + 268   + 269   + 585   + 906   + 232   + 104   + 129
  1 156   1 011   1 036   1 852   434   226   499

Find the sum

Complete all the activities (Addition).

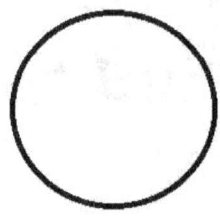

SCORE

| 1. | 622 + 991 | 2. | 450 + 228 | 3. | 985 + 138 | 4. | 730 + 617 | 5. | 437 + 374 | 6. | 677 + 789 | 7. | 307 + 703 |

1.   622 + 991
2.   450 + 228
3.   985 + 138
4.   730 + 617
5.   437 + 374
6.   677 + 789
7.   307 + 703

8.   918 + 872
9.   671 + 487
10.   691 + 653
11.   925 + 997
12.   412 + 937
13.   840 + 405
14.   302 + 187

15.   244 + 232
16.   767 + 967
17.   706 + 460
18.   515 + 696
19.   986 + 975
20.   122 + 922
21.   412 + 370

22.   831 + 467
23.   591 + 661
24.   378 + 343
25.   987 + 329
26.   561 + 634
27.   413 + 717
28.   203 + 624

29.   679 + 512
30.   387 + 680
31.   834 + 642
32.   473 + 305
33.   607 + 709
34.   183 + 846
35.   289 + 322

36.   885 + 694
37.   465 + 767
38.   341 + 655
39.   266 + 943
40.   447 + 279
41.   127 + 959
42.   199 + 914

Name:................................   Date:................................

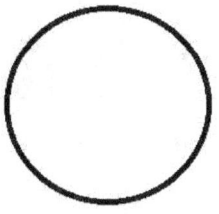

SCORE

Find the sum

Complete all the activities (Addition).

1.	622 + 991	2.	450 + 228	3.	985 + 138	4.	730 + 617	5.	437 + 374	6.	677 + 789	7.	307 + 703
	1 613		678		1 123		1 347		811		1 466		1 010

8.	918 + 872	9.	671 + 487	10.	691 + 653	11.	925 + 997	12.	412 + 937	13.	840 + 405	14.	302 + 187
	1 790		1 158		1 344		1 922		1 349		1 245		489

15.	244 + 232	16.	767 + 967	17.	706 + 460	18.	515 + 696	19.	986 + 975	20.	122 + 922	21.	412 + 370
	476		1 734		1 166		1 211		1 961		1 044		782

22.	831 + 467	23.	591 + 661	24.	378 + 343	25.	987 + 329	26.	561 + 634	27.	413 + 717	28.	203 + 624
	1 298		1 252		721		1 316		1 195		1 130		827

29.	679 + 512	30.	387 + 680	31.	834 + 642	32.	473 + 305	33.	607 + 709	34.	183 + 846	35.	289 + 322
	1 191		1 067		1 476		778		1 316		1 029		611

36.	885 + 694	37.	465 + 767	38.	341 + 655	39.	266 + 943	40.	447 + 279	41.	127 + 959	42.	199 + 914
	1 579		1 232		996		1 209		726		1 086		1 113

Find the sum

Complete all the activities (Addition).

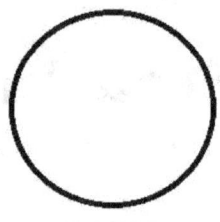

SCORE

1. 670 + 233	2. 654 + 988	3. 357 + 889	4. 116 + 574	5. 689 + 441	6. 188 + 769	7. 896 + 626
8. 860 + 881	9. 578 + 531	10. 154 + 765	11. 443 + 865	12. 269 + 645	13. 476 + 547	14. 458 + 735
15. 165 + 208	16. 653 + 875	17. 543 + 558	18. 633 + 474	19. 155 + 193	20. 589 + 398	21. 306 + 411
22. 106 + 124	23. 679 + 571	24. 757 + 639	25. 183 + 565	26. 700 + 595	27. 905 + 433	28. 570 + 804
29. 963 + 110	30. 613 + 809	31. 532 + 380	32. 509 + 581	33. 787 + 312	34. 325 + 438	35. 992 + 486
36. 826 + 485	37. 941 + 754	38. 559 + 133	39. 860 + 245	40. 965 + 904	41. 789 + 817	42. 652 + 599

Find the sum

Complete all the activities (Addition).

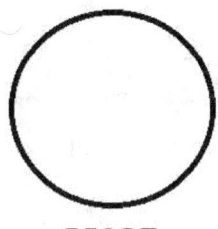

SCORE

1. 670 + 233 903	2. 654 + 988 1 642	3. 357 + 889 1 246	4. 116 + 574 690	5. 689 + 441 1 130	6. 188 + 769 957	7. 896 + 626 1 522
8. 860 + 881 1 741	9. 578 + 531 1 109	10. 154 + 765 919	11. 443 + 865 1 308	12. 269 + 645 914	13. 476 + 547 1 023	14. 458 + 735 1 193
15. 165 + 208 373	16. 653 + 875 1 528	17. 543 + 558 1 101	18. 633 + 474 1 107	19. 155 + 193 348	20. 589 + 398 987	21. 306 + 411 717
22. 106 + 124 230	23. 679 + 571 1 250	24. 757 + 639 1 396	25. 183 + 565 748	26. 700 + 595 1 295	27. 905 + 433 1 338	28. 570 + 804 1 374
29. 963 + 110 1 073	30. 613 + 809 1 422	31. 532 + 380 912	32. 509 + 581 1 090	33. 787 + 312 1 099	34. 325 + 438 763	35. 992 + 486 1 478
36. 826 + 485 1 311	37. 941 + 754 1 695	38. 559 + 133 692	39. 860 + 245 1 105	40. 965 + 904 1 869	41. 789 + 817 1 606	42. 652 + 599 1 251

Find the sum

Complete all the activities (Addition).

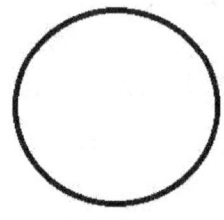

SCORE

1.   507
  + 158
	2. 969
	  + 210
		3. 987
		  + 828
			4. 510
			  + 145
				5. 580
				  + 779
					6. 866
					  + 925
						7. 901
						  + 524

8.   335
  + 467
	9. 346
	  + 419
		10. 647
		  + 696
			11. 738
			  + 315
				12. 653
				  + 251
					13. 873
					  + 538
						14. 943
						  + 818

15.   144
  + 708
	16. 120
	  + 713
		17. 727
		  + 908
			18. 174
			  + 729
				19. 343
				  + 215
					20. 450
					  + 264
						21. 975
						  + 781

22.   904
  + 262
	23. 888
	  + 634
		24. 231
		  + 436
			25. 909
			  + 598
				26. 658
				  + 470
					27. 271
					  + 954
						28. 860
						  + 292

29.   504
  + 818
	30. 336
	  + 549
		31. 884
		  + 691
			32. 101
			  + 871
				33. 759
				  + 252
					34. 297
					  + 991
						35. 449
						  + 605

36.   224
  + 230
	37. 968
	  + 483
		38. 720
		  + 956
			39. 410
			  + 490
				40. 362
				  + 339
					41. 407
					  + 607
						42. 908
						  + 545

Find the sum

Complete all the activities (Addition).

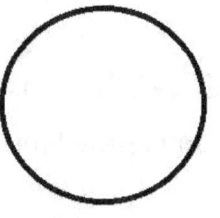

SCORE

| 1. | 507
 + 158
 665 | 2. | 969
 + 210
 1 179 | 3. | 987
 + 828
 1 815 | 4. | 510
 + 145
 655 | 5. | 580
 + 779
 1 359 | 6. | 866
 + 925
 1 791 | 7. | 901
 + 524
 1 425 |

8. 335   9. 346   10. 647   11. 738   12. 653   13. 873   14. 943
   + 467   + 419   + 696   + 315   + 251   + 538   + 818
    802   765   1 343   1 053   904   1 411   1 761

15. 144   16. 120   17. 727   18. 174   19. 343   20. 450   21. 975
   + 708   + 713   + 908   + 729   + 215   + 264   + 781
    852   833   1 635   903   558   714   1 756

22. 904   23. 888   24. 231   25. 909   26. 658   27. 271   28. 860
   + 262   + 634   + 436   + 598   + 470   + 954   + 292
  1 166   1 522   667   1 507   1 128   1 225   1 152

29. 504   30. 336   31. 884   32. 101   33. 759   34. 297   35. 449
   + 818   + 549   + 691   + 871   + 252   + 991   + 605
  1 322   885   1 575   972   1 011   1 288   1 054

36. 224   37. 968   38. 720   39. 410   40. 362   41. 407   42. 908
   + 230   + 483   + 956   + 490   + 339   + 607   + 545
    454   1 451   1 676   900   701   1 014   1 453

Name:................................   Date:................................

Find the sum

Complete all the activities (Addition).

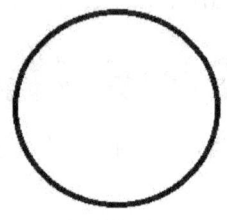

1. 150 + 649	2. 543 + 772	3. 148 + 881	4. 325 + 523	5. 878 + 900	6. 379 + 384	7. 781 + 133
8. 503 + 736	9. 436 + 239	10. 986 + 404	11. 474 + 527	12. 501 + 787	13. 334 + 903	14. 881 + 114
15. 556 + 996	16. 243 + 377	17. 782 + 554	18. 821 + 210	19. 708 + 525	20. 943 + 567	21. 380 + 700
22. 351 + 167	23. 204 + 436	24. 275 + 377	25. 987 + 625	26. 902 + 268	27. 620 + 604	28. 172 + 622
29. 247 + 755	30. 538 + 244	31. 999 + 311	32. 966 + 881	33. 758 + 101	34. 601 + 326	35. 311 + 393
36. 127 + 910	37. 294 + 519	38. 124 + 773	39. 276 + 434	40. 682 + 898	41. 728 + 604	42. 126 + 930

Name:................................. Date:.............................

Find the sum

Complete all the activities (Addition).

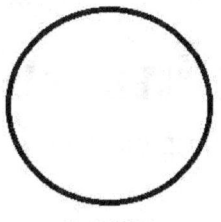

SCORE

| 1. | 150
+ 649
799 | 2. | 543
+ 772
1 315 | 3. | 148
+ 881
1 029 | 4. | 325
+ 523
848 | 5. | 878
+ 900
1 778 | 6. | 379
+ 384
763 | 7. | 781
+ 133
914 |

| 8. | 503
+ 736
1 239 | 9. | 436
+ 239
675 | 10. | 986
+ 404
1 390 | 11. | 474
+ 527
1 001 | 12. | 501
+ 787
1 288 | 13. | 334
+ 903
1 237 | 14. | 881
+ 114
995 |

| 15. | 556
+ 996
1 552 | 16. | 243
+ 377
620 | 17. | 782
+ 554
1 336 | 18. | 821
+ 210
1 031 | 19. | 708
+ 525
1 233 | 20. | 943
+ 567
1 510 | 21. | 380
+ 700
1 080 |

| 22. | 351
+ 167
518 | 23. | 204
+ 436
640 | 24. | 275
+ 377
652 | 25. | 987
+ 625
1 612 | 26. | 902
+ 268
1 170 | 27. | 620
+ 604
1 224 | 28. | 172
+ 622
794 |

| 29. | 247
+ 755
1 002 | 30. | 538
+ 244
782 | 31. | 999
+ 311
1 310 | 32. | 966
+ 881
1 847 | 33. | 758
+ 101
859 | 34. | 601
+ 326
927 | 35. | 311
+ 393
704 |

| 36. | 127
+ 910
1 037 | 37. | 294
+ 519
813 | 38. | 124
+ 773
897 | 39. | 276
+ 434
710 | 40. | 682
+ 898
1 580 | 41. | 728
+ 604
1 332 | 42. | 126
+ 930
1 056 |

Find the sum

Complete all the activities (Addition).

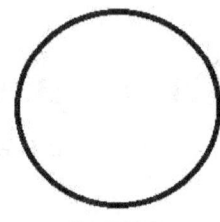

SCORE

1.   194
  + 974
	2. 756
	  + 279
		3. 483
		  + 299
			4. 962
			  + 952
				5. 986
				  + 799
					6. 567
					  + 168
						7. 733
						  + 469

8.   408
  + 203
	9. 834
	  + 948
		10. 165
		  + 694
			11. 598
			  + 991
				12. 972
				  + 201
					13. 579
					  + 398
						14. 446
						  + 812

15.   173
  + 639
	16. 515
	  + 494
		17. 286
		  + 136
			18. 787
			  + 783
				19. 344
				  + 423
					20. 843
					  + 113
						21. 430
						  + 174

22.   923
  + 494
	23. 951
	  + 955
		24. 140
		  + 806
			25. 126
			  + 579
				26. 164
				  + 779
					27. 539
					  + 431
						28. 432
						  + 202

29.   101
  + 655
	30. 698
	  + 185
		31. 852
		  + 249
			32. 880
			  + 447
				33. 384
				  + 173
					34. 882
					  + 697
						35. 681
						  + 624

36.   858
  + 511
	37. 205
	  + 153
		38. 364
		  + 525
			39. 167
			  + 579
				40. 250
				  + 623
					41. 357
					  + 639
						42. 787
						  + 592

Find the sum

Complete all the activities (Addition).

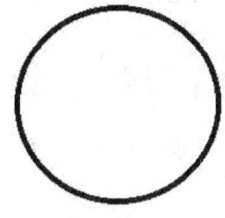

SCORE

| 1. | 194
+ 974
1 168 | 2. | 756
+ 279
1 035 | 3. | 483
+ 299
782 | 4. | 962
+ 952
1 914 | 5. | 986
+ 799
1 785 | 6. | 567
+ 168
735 | 7. | 733
+ 469
1 202 |

| 8. | 408
+ 203
611 | 9. | 834
+ 948
1 782 | 10. | 165
+ 694
859 | 11. | 598
+ 991
1 589 | 12. | 972
+ 201
1 173 | 13. | 579
+ 398
977 | 14. | 446
+ 812
1 258 |

| 15. | 173
+ 639
812 | 16. | 515
+ 494
1 009 | 17. | 286
+ 136
422 | 18. | 787
+ 783
1 570 | 19. | 344
+ 423
767 | 20. | 843
+ 113
956 | 21. | 430
+ 174
604 |

| 22. | 923
+ 494
1 417 | 23. | 951
+ 955
1 906 | 24. | 140
+ 806
946 | 25. | 126
+ 579
705 | 26. | 164
+ 779
943 | 27. | 539
+ 431
970 | 28. | 432
+ 202
634 |

| 29. | 101
+ 655
756 | 30. | 698
+ 185
883 | 31. | 852
+ 249
1 101 | 32. | 880
+ 447
1 327 | 33. | 384
+ 173
557 | 34. | 882
+ 697
1 579 | 35. | 681
+ 624
1 305 |

| 36. | 858
+ 511
1 369 | 37. | 205
+ 153
358 | 38. | 364
+ 525
889 | 39. | 167
+ 579
746 | 40. | 250
+ 623
873 | 41. | 357
+ 639
996 | 42. | 787
+ 592
1 379 |

Find the sum

Complete all the activities (Addition).

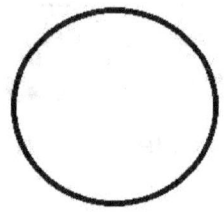

SCORE

1. 575 + 735	2. 954 + 854	3. 703 + 685	4. 782 + 688	5. 723 + 758	6. 260 + 281	7. 293 + 576
8. 210 + 878	9. 173 + 442	10. 224 + 500	11. 720 + 652	12. 674 + 486	13. 166 + 906	14. 135 + 522
15. 506 + 175	16. 827 + 313	17. 462 + 388	18. 871 + 648	19. 217 + 128	20. 993 + 996	21. 842 + 270
22. 787 + 446	23. 892 + 184	24. 332 + 590	25. 469 + 371	26. 550 + 914	27. 515 + 386	28. 871 + 922
29. 162 + 568	30. 788 + 796	31. 650 + 360	32. 282 + 648	33. 930 + 946	34. 818 + 904	35. 704 + 561
36. 761 + 145	37. 448 + 227	38. 658 + 550	39. 864 + 327	40. 593 + 981	41. 675 + 544	42. 153 + 235

Find the sum

Complete all the activities (Addition).

Name:.............................. Date:..............................

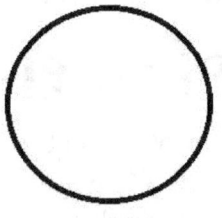

SCORE

1. 575 + 735 1 310	2. 954 + 854 1 808	3. 703 + 685 1 388	4. 782 + 688 1 470	5. 723 + 758 1 481	6. 260 + 281 541	7. 293 + 576 869
8. 210 + 878 1 088	9. 173 + 442 615	10. 224 + 500 724	11. 720 + 652 1 372	12. 674 + 486 1 160	13. 166 + 906 1 072	14. 135 + 522 657
15. 506 + 175 681	16. 827 + 313 1 140	17. 462 + 388 850	18. 871 + 648 1 519	19. 217 + 128 345	20. 993 + 996 1 989	21. 842 + 270 1 112
22. 787 + 446 1 233	23. 892 + 184 1 076	24. 332 + 590 922	25. 469 + 371 840	26. 550 + 914 1 464	27. 515 + 386 901	28. 871 + 922 1 793
29. 162 + 568 730	30. 788 + 796 1 584	31. 650 + 360 1 010	32. 282 + 648 930	33. 930 + 946 1 876	34. 818 + 904 1 722	35. 704 + 561 1 265
36. 761 + 145 906	37. 448 + 227 675	38. 658 + 550 1 208	39. 864 + 327 1 191	40. 593 + 981 1 574	41. 675 + 544 1 219	42. 153 + 235 388

Find the sum

Complete all the activities (Addition).

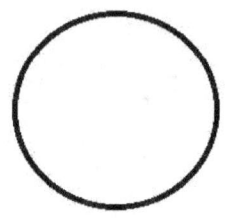

SCORE

| 1. 624 + 912 | 2. 796 + 175 | 3. 516 + 948 | 4. 586 + 648 | 5. 339 + 298 | 6. 446 + 582 | 7. 987 + 238 |

1. 624
  + 912
  ----

2. 796
  + 175
  ----

3. 516
  + 948
  ----

4. 586
  + 648
  ----

5. 339
  + 298
  ----

6. 446
  + 582
  ----

7. 987
  + 238
  ----

8. 515
  + 946
  ----

9. 365
  + 426
  ----

10. 885
  + 907
  ----

11. 609
  + 505
  ----

12. 566
  + 388
  ----

13. 728
  + 421
  ----

14. 158
  + 834
  ----

15. 974
  + 806
  ----

16. 221
  + 206
  ----

17. 307
  + 148
  ----

18. 124
  + 119
  ----

19. 249
  + 743
  ----

20. 489
  + 755
  ----

21. 974
  + 212
  ----

22. 805
  + 250
  ----

23. 780
  + 176
  ----

24. 563
  + 726
  ----

25. 811
  + 646
  ----

26. 717
  + 348
  ----

27. 658
  + 245
  ----

28. 624
  + 220
  ----

29. 445
  + 509
  ----

30. 447
  + 195
  ----

31. 891
  + 751
  ----

32. 673
  + 532
  ----

33. 177
  + 340
  ----

34. 374
  + 939
  ----

35. 215
  + 496
  ----

36. 796
  + 621
  ----

37. 937
  + 230
  ----

38. 932
  + 497
  ----

39. 698
  + 416
  ----

40. 606
  + 652
  ----

41. 857
  + 431
  ----

42. 122
  + 428

Find the sum

Complete all the activities (Addition).

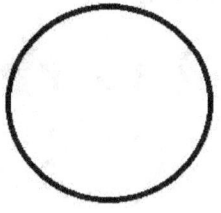

SCORE

| 1. | 624
+ 912
1 536 | 2. | 796
+ 175
971 | 3. | 516
+ 948
1 464 | 4. | 586
+ 648
1 234 | 5. | 339
+ 298
637 | 6. | 446
+ 582
1 028 | 7. | 987
+ 238
1 225 |

| 8. | 515
+ 946
1 461 | 9. | 365
+ 426
791 | 10. | 885
+ 907
1 792 | 11. | 609
+ 505
1 114 | 12. | 566
+ 388
954 | 13. | 728
+ 421
1 149 | 14. | 158
+ 834
992 |

| 15. | 974
+ 806
1 780 | 16. | 221
+ 206
427 | 17. | 307
+ 148
455 | 18. | 124
+ 119
243 | 19. | 249
+ 743
992 | 20. | 489
+ 755
1 244 | 21. | 974
+ 212
1 186 |

| 22. | 805
+ 250
1 055 | 23. | 780
+ 176
956 | 24. | 563
+ 726
1 289 | 25. | 811
+ 646
1 457 | 26. | 717
+ 348
1 065 | 27. | 658
+ 245
903 | 28. | 624
+ 220
844 |

| 29. | 445
+ 509
954 | 30. | 447
+ 195
642 | 31. | 891
+ 751
1 642 | 32. | 673
+ 532
1 205 | 33. | 177
+ 340
517 | 34. | 374
+ 939
1 313 | 35. | 215
+ 496
711 |

| 36. | 796
+ 621
1 417 | 37. | 937
+ 230
1 167 | 38. | 932
+ 497
1 429 | 39. | 698
+ 416
1 114 | 40. | 606
+ 652
1 258 | 41. | 857
+ 431
1 288 | 42. | 122
+ 428
550 |

Find the sum

Complete all the activities (Addition).

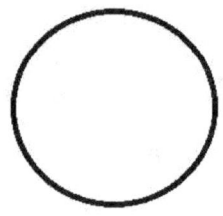

1. 189
    + 872
    ----

2. 547
    + 162
    ----

3. 522
    + 903
    ----

4. 672
    + 740
    ----

5. 396
    + 394
    ----

6. 863
    + 323
    ----

7. 539
    + 373
    ----

8. 173
    + 540
    ----

9. 801
    + 623
    ----

10. 611
     + 415
     ----

11. 849
     + 196
     ----

12. 410
     + 528
     ----

13. 312
     + 228
     ----

14. 520
     + 129
     ----

15. 901
     + 794
     ----

16. 337
     + 560
     ----

17. 729
     + 344
     ----

18. 443
     + 304
     ----

19. 793
     + 312
     ----

20. 517
     + 310
     ----

21. 823
     + 157
     ----

22. 325
     + 643
     ----

23. 366
     + 709
     ----

24. 957
     + 425
     ----

25. 343
     + 166
     ----

26. 850
     + 334
     ----

27. 818
     + 706
     ----

28. 578
     + 993
     ----

29. 561
     + 118
     ----

30. 850
     + 395
     ----

31. 483
     + 643
     ----

32. 837
     + 601
     ----

33. 157
     + 403
     ----

34. 593
     + 295
     ----

35. 929
     + 868
     ----

36. 747
     + 589
     ----

37. 206
     + 401
     ----

38. 591
     + 792
     ----

39. 870
     + 766
     ----

40. 263
     + 602
     ----

41. 270
     + 195
     ----

42. 786
     + 835

Find the sum

Name:..................................... Date:.....................................

Complete all the activities (Addition).

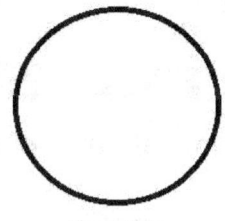

SCORE

| 1. | 189
+ 872
1 061 | 2. | 547
+ 162
709 | 3. | 522
+ 903
1 425 | 4. | 672
+ 740
1 412 | 5. | 396
+ 394
790 | 6. | 863
+ 323
1 186 | 7. | 539
+ 373
912 |

| 8. | 173
+ 540
713 | 9. | 801
+ 623
1 424 | 10. | 611
+ 415
1 026 | 11. | 849
+ 196
1 045 | 12. | 410
+ 528
938 | 13. | 312
+ 228
540 | 14. | 520
+ 129
649 |

| 15. | 901
+ 794
1 695 | 16. | 337
+ 560
897 | 17. | 729
+ 344
1 073 | 18. | 443
+ 304
747 | 19. | 793
+ 312
1 105 | 20. | 517
+ 310
827 | 21. | 823
+ 157
980 |

| 22. | 325
+ 643
968 | 23. | 366
+ 709
1 075 | 24. | 957
+ 425
1 382 | 25. | 343
+ 166
509 | 26. | 850
+ 334
1 184 | 27. | 818
+ 706
1 524 | 28. | 578
+ 993
1 571 |

| 29. | 561
+ 118
679 | 30. | 850
+ 395
1 245 | 31. | 483
+ 643
1 126 | 32. | 837
+ 601
1 438 | 33. | 157
+ 403
560 | 34. | 593
+ 295
888 | 35. | 929
+ 868
1 797 |

| 36. | 747
+ 589
1 336 | 37. | 206
+ 401
607 | 38. | 591
+ 792
1 383 | 39. | 870
+ 766
1 636 | 40. | 263
+ 602
865 | 41. | 270
+ 195
465 | 42. | 786
+ 835
1 621 |

Find the sum

Complete all the activities (Addition).

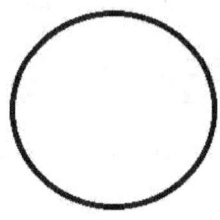

SCORE

1. 943 + 103	2. 741 + 123	3. 765 + 450	4. 742 + 774	5. 461 + 222	6. 826 + 280	7. 318 + 722
8. 586 + 961	9. 216 + 762	10. 496 + 474	11. 418 + 360	12. 228 + 517	13. 215 + 785	14. 223 + 794
15. 854 + 348	16. 225 + 388	17. 856 + 102	18. 657 + 154	19. 661 + 570	20. 529 + 579	21. 226 + 971
22. 228 + 983	23. 574 + 411	24. 574 + 247	25. 280 + 300	26. 485 + 506	27. 698 + 412	28. 517 + 683
29. 874 + 324	30. 907 + 979	31. 327 + 948	32. 919 + 213	33. 996 + 538	34. 423 + 316	35. 166 + 532
36. 400 + 645	37. 334 + 915	38. 348 + 446	39. 232 + 244	40. 943 + 803	41. 311 + 974	42. 614 + 654

Find the sum

Name:.......................... Date:..........................

Complete all the activities (Addition).

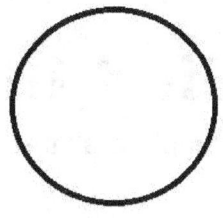

SCORE

| 1. | 943
+ 103
1 046 | 2. | 741
+ 123
864 | 3. | 765
+ 450
1 215 | 4. | 742
+ 774
1 516 | 5. | 461
+ 222
683 | 6. | 826
+ 280
1 106 | 7. | 318
+ 722
1 040 |

| 8. | 586
+ 961
1 547 | 9. | 216
+ 762
978 | 10. | 496
+ 474
970 | 11. | 418
+ 360
778 | 12. | 228
+ 517
745 | 13. | 215
+ 785
1 000 | 14. | 223
+ 794
1 017 |

| 15. | 854
+ 348
1 202 | 16. | 225
+ 388
613 | 17. | 856
+ 102
958 | 18. | 657
+ 154
811 | 19. | 661
+ 570
1 231 | 20. | 529
+ 579
1 108 | 21. | 226
+ 971
1 197 |

| 22. | 228
+ 983
1 211 | 23. | 574
+ 411
985 | 24. | 574
+ 247
821 | 25. | 280
+ 300
580 | 26. | 485
+ 506
991 | 27. | 698
+ 412
1 110 | 28. | 517
+ 683
1 200 |

| 29. | 874
+ 324
1 198 | 30. | 907
+ 979
1 886 | 31. | 327
+ 948
1 275 | 32. | 919
+ 213
1 132 | 33. | 996
+ 538
1 534 | 34. | 423
+ 316
739 | 35. | 166
+ 532
698 |

| 36. | 400
+ 645
1 045 | 37. | 334
+ 915
1 249 | 38. | 348
+ 446
794 | 39. | 232
+ 244
476 | 40. | 943
+ 803
1 746 | 41. | 311
+ 974
1 285 | 42. | 614
+ 654
1 268 |

Find the sum

Complete all the activities (Addition).

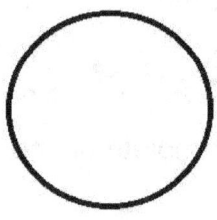

SCORE

1. 861 + 717	2. 688 + 544	3. 584 + 973	4. 280 + 437	5. 662 + 549	6. 352 + 277	7. 152 + 115
8. 986 + 135	9. 988 + 357	10. 182 + 747	11. 899 + 839	12. 273 + 782	13. 165 + 171	14. 550 + 730
15. 380 + 204	16. 768 + 352	17. 434 + 365	18. 351 + 379	19. 849 + 878	20. 732 + 192	21. 352 + 962
22. 982 + 784	23. 663 + 171	24. 394 + 218	25. 257 + 907	26. 814 + 953	27. 893 + 865	28. 754 + 122
29. 545 + 905	30. 479 + 513	31. 522 + 549	32. 157 + 558	33. 780 + 100	34. 842 + 786	35. 763 + 707
36. 285 + 210	37. 625 + 989	38. 495 + 936	39. 856 + 488	40. 219 + 993	41. 776 + 433	42. 573 + 472

Find the sum

Complete all the activities (Addition).

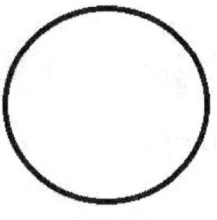

SCORE

Name:................................ Date:.................................

| 1. | 861 + 717 = 1 578 | 2. | 688 + 544 = 1 232 | 3. | 584 + 973 = 1 557 | 4. | 280 + 437 = 717 | 5. | 662 + 549 = 1 211 | 6. | 352 + 277 = 629 | 7. | 152 + 115 = 267 |

1.   861
   + 717
   1 578

2.   688
   + 544
   1 232

3.   584
   + 973
   1 557

4.   280
   + 437
   717

5.   662
   + 549
   1 211

6.   352
   + 277
   629

7.   152
   + 115
   267

8.   986
   + 135
   1 121

9.   988
   + 357
   1 345

10.   182
   + 747
   929

11.   899
   + 839
   1 738

12.   273
   + 782
   1 055

13.   165
   + 171
   336

14.   550
   + 730
   1 280

15.   380
   + 204
   584

16.   768
   + 352
   1 120

17.   434
   + 365
   799

18.   351
   + 379
   730

19.   849
   + 878
   1 727

20.   732
   + 192
   924

21.   352
   + 962
   1 314

22.   982
   + 784
   1 766

23.   663
   + 171
   834

24.   394
   + 218
   612

25.   257
   + 907
   1 164

26.   814
   + 953
   1 767

27.   893
   + 865
   1 758

28.   754
   + 122
   876

29.   545
   + 905
   1 450

30.   479
   + 513
   992

31.   522
   + 549
   1 071

32.   157
   + 558
   715

33.   780
   + 100
   880

34.   842
   + 786
   1 628

35.   763
   + 707
   1 470

36.   285
   + 210
   495

37.   625
   + 989
   1 614

38.   495
   + 936
   1 431

39.   856
   + 488
   1 344

40.   219
   + 993
   1 212

41.   776
   + 433
   1 209

42.   573
   + 472
   1 045

Name:..................................   Date:..................................

Find the sum

Complete all the activities (Addition).

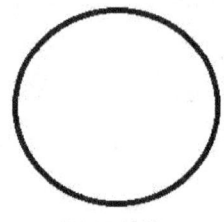

SCORE

1. 389 + 814	2. 458 + 568	3. 610 + 328	4. 833 + 943	5. 498 + 319	6. 669 + 803	7. 905 + 524
8. 104 + 663	9. 328 + 607	10. 454 + 549	11. 722 + 704	12. 428 + 903	13. 471 + 816	14. 265 + 157
15. 787 + 485	16. 736 + 190	17. 975 + 684	18. 269 + 964	19. 641 + 374	20. 276 + 534	21. 873 + 342
22. 711 + 406	23. 420 + 691	24. 132 + 433	25. 483 + 555	26. 765 + 269	27. 978 + 124	28. 748 + 684
29. 487 + 554	30. 197 + 793	31. 163 + 776	32. 256 + 113	33. 581 + 579	34. 826 + 320	35. 355 + 139
36. 122 + 952	37. 431 + 919	38. 675 + 481	39. 992 + 121	40. 583 + 779	41. 785 + 832	42. 673 + 332

Find the sum

Name:............................... Date:...............................

Complete all the activities (Addition).

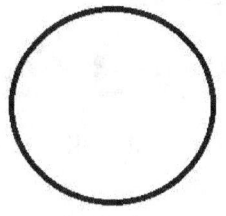

SCORE

1. 389 + 814 1 203	2. 458 + 568 1 026	3. 610 + 328 938	4. 833 + 943 1 776	5. 498 + 319 817	6. 669 + 803 1 472	7. 905 + 524 1 429
8. 104 + 663 767	9. 328 + 607 935	10. 454 + 549 1 003	11. 722 + 704 1 426	12. 428 + 903 1 331	13. 471 + 816 1 287	14. 265 + 157 422
15. 787 + 485 1 272	16. 736 + 190 926	17. 975 + 684 1 659	18. 269 + 964 1 233	19. 641 + 374 1 015	20. 276 + 534 810	21. 873 + 342 1 215
22. 711 + 406 1 117	23. 420 + 691 1 111	24. 132 + 433 565	25. 483 + 555 1 038	26. 765 + 269 1 034	27. 978 + 124 1 102	28. 748 + 684 1 432
29. 487 + 554 1 041	30. 197 + 793 990	31. 163 + 776 939	32. 256 + 113 369	33. 581 + 579 1 160	34. 826 + 320 1 146	35. 355 + 139 494
36. 122 + 952 1 074	37. 431 + 919 1 350	38. 675 + 481 1 156	39. 992 + 121 1 113	40. 583 + 779 1 362	41. 785 + 832 1 617	42. 673 + 332 1 005

Find the sum

Complete all the activities (Addition).

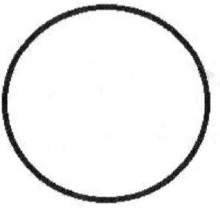

SCORE

1. 719 + 568	2. 211 + 951	3. 935 + 584	4. 578 + 187	5. 277 + 482	6. 471 + 197	7. 186 + 698
8. 824 + 840	9. 864 + 810	10. 178 + 444	11. 533 + 812	12. 836 + 159	13. 448 + 368	14. 691 + 465
15. 971 + 418	16. 108 + 331	17. 203 + 922	18. 393 + 114	19. 468 + 874	20. 385 + 770	21. 561 + 934
22. 775 + 949	23. 548 + 577	24. 395 + 395	25. 296 + 741	26. 914 + 974	27. 754 + 329	28. 799 + 906
29. 730 + 152	30. 603 + 667	31. 891 + 212	32. 215 + 859	33. 474 + 240	34. 994 + 263	35. 405 + 810
36. 473 + 988	37. 526 + 407	38. 882 + 598	39. 137 + 887	40. 610 + 870	41. 972 + 397	42. 269 + 330

Find the sum

Name:...................................... Date:......................................

Complete all the activities (Addition).

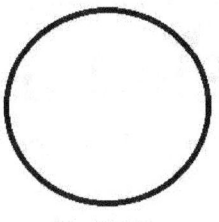

SCORE

1. 719 + 568 1 287	2. 211 + 951 1 162	3. 935 + 584 1 519	4. 578 + 187 765	5. 277 + 482 759	6. 471 + 197 668	7. 186 + 698 884
8. 824 + 840 1 664	9. 864 + 810 1 674	10. 178 + 444 622	11. 533 + 812 1 345	12. 836 + 159 995	13. 448 + 368 816	14. 691 + 465 1 156
15. 971 + 418 1 389	16. 108 + 331 439	17. 203 + 922 1 125	18. 393 + 114 507	19. 468 + 874 1 342	20. 385 + 770 1 155	21. 561 + 934 1 495
22. 775 + 949 1 724	23. 548 + 577 1 125	24. 395 + 395 790	25. 296 + 741 1 037	26. 914 + 974 1 888	27. 754 + 329 1 083	28. 799 + 906 1 705
29. 730 + 152 882	30. 603 + 667 1 270	31. 891 + 212 1 103	32. 215 + 859 1 074	33. 474 + 240 714	34. 994 + 263 1 257	35. 405 + 810 1 215
36. 473 + 988 1 461	37. 526 + 407 933	38. 882 + 598 1 480	39. 137 + 887 1 024	40. 610 + 870 1 480	41. 972 + 397 1 369	42. 269 + 330 599

Name:.................................. Date:..................................

Find the sum

Complete all the activities (Addition).

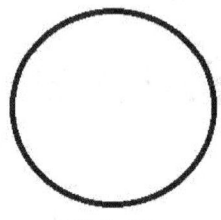

| 1. 503
+ 133 | 2. 576
+ 872 | 3. 224
+ 486 | 4. 382
+ 416 | 5. 372
+ 716 | 6. 878
+ 448 | 7. 819
+ 916 |

1. 503
    + 133
    ----

2. 576
    + 872
    ----

3. 224
    + 486
    ----

4. 382
    + 416
    ----

5. 372
    + 716
    ----

6. 878
    + 448
    ----

7. 819
    + 916
    ----

8. 575
    + 408
    ----

9. 583
    + 232
    ----

10. 989
    + 485
    ----

11. 369
    + 925
    ----

12. 548
    + 447
    ----

13. 446
    + 301
    ----

14. 870
    + 843
    ----

15. 541
    + 732
    ----

16. 694
    + 595
    ----

17. 942
    + 884
    ----

18. 456
    + 441
    ----

19. 663
    + 222
    ----

20. 139
    + 918
    ----

21. 528
    + 157
    ----

22. 717
    + 808
    ----

23. 528
    + 534
    ----

24. 778
    + 220
    ----

25. 412
    + 302
    ----

26. 387
    + 840
    ----

27. 407
    + 783
    ----

28. 960
    + 577
    ----

29. 584
    + 229
    ----

30. 526
    + 922
    ----

31. 615
    + 539
    ----

32. 873
    + 639
    ----

33. 122
    + 383
    ----

34. 505
    + 934
    ----

35. 560
    + 234
    ----

36. 744
    + 461
    ----

37. 194
    + 171
    ----

38. 956
    + 312
    ----

39. 308
    + 739
    ----

40. 918
    + 912
    ----

41. 522
    + 749
    ----

42. 531
    + 339

Find the sum

Complete all the activities (Addition).

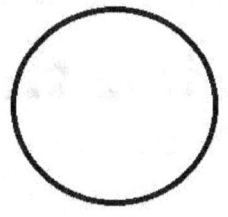

SCORE

| 1. | 503
+ 133
636 | 2. | 576
+ 872
1 448 | 3. | 224
+ 486
710 | 4. | 382
+ 416
798 | 5. | 372
+ 716
1 088 | 6. | 878
+ 448
1 326 | 7. | 819
+ 916
1 735 |

| 8. | 575
+ 408
983 | 9. | 583
+ 232
815 | 10. | 989
+ 485
1 474 | 11. | 369
+ 925
1 294 | 12. | 548
+ 447
995 | 13. | 446
+ 301
747 | 14. | 870
+ 843
1 713 |

| 15. | 541
+ 732
1 273 | 16. | 694
+ 595
1 289 | 17. | 942
+ 884
1 826 | 18. | 456
+ 441
897 | 19. | 663
+ 222
885 | 20. | 139
+ 918
1 057 | 21. | 528
+ 157
685 |

| 22. | 717
+ 808
1 525 | 23. | 528
+ 534
1 062 | 24. | 778
+ 220
998 | 25. | 412
+ 302
714 | 26. | 387
+ 840
1 227 | 27. | 407
+ 783
1 190 | 28. | 960
+ 577
1 537 |

| 29. | 584
+ 229
813 | 30. | 526
+ 922
1 448 | 31. | 615
+ 539
1 154 | 32. | 873
+ 639
1 512 | 33. | 122
+ 383
505 | 34. | 505
+ 934
1 439 | 35. | 560
+ 234
794 |

| 36. | 744
+ 461
1 205 | 37. | 194
+ 171
365 | 38. | 956
+ 312
1 268 | 39. | 308
+ 739
1 047 | 40. | 918
+ 912
1 830 | 41. | 522
+ 749
1 271 | 42. | 531
+ 339
870 |

Find the sum

Complete all the activities (Addition).

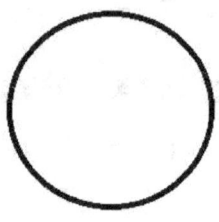

SCORE

1.	444 + 854	2.	288 + 259	3.	923 + 878	4.	856 + 533	5.	695 + 216	6.	403 + 988	7.	283 + 178
8.	786 + 245	9.	128 + 843	10.	660 + 278	11.	557 + 925	12.	746 + 336	13.	826 + 748	14.	369 + 844
15.	310 + 613	16.	499 + 865	17.	108 + 634	18.	295 + 267	19.	195 + 346	20.	110 + 680	21.	377 + 311
22.	765 + 977	23.	766 + 967	24.	129 + 231	25.	952 + 773	26.	883 + 679	27.	190 + 353	28.	781 + 518
29.	207 + 951	30.	433 + 275	31.	609 + 113	32.	323 + 181	33.	715 + 537	34.	980 + 647	35.	882 + 619
36.	150 + 782	37.	833 + 978	38.	900 + 346	39.	260 + 768	40.	569 + 857	41.	719 + 238	42.	891 + 536

Find the sum

Complete all the activities (Addition).

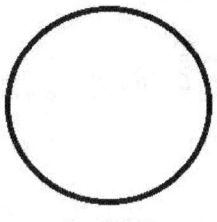

SCORE

1. 444 + 854 1 298	2. 288 + 259 547	3. 923 + 878 1 801	4. 856 + 533 1 389	5. 695 + 216 911	6. 403 + 988 1 391	7. 283 + 178 461
8. 786 + 245 1 031	9. 128 + 843 971	10. 660 + 278 938	11. 557 + 925 1 482	12. 746 + 336 1 082	13. 826 + 748 1 574	14. 369 + 844 1 213
15. 310 + 613 923	16. 499 + 865 1 364	17. 108 + 634 742	18. 295 + 267 562	19. 195 + 346 541	20. 110 + 680 790	21. 377 + 311 688
22. 765 + 977 1 742	23. 766 + 967 1 733	24. 129 + 231 360	25. 952 + 773 1 725	26. 883 + 679 1 562	27. 190 + 353 543	28. 781 + 518 1 299
29. 207 + 951 1 158	30. 433 + 275 708	31. 609 + 113 722	32. 323 + 181 504	33. 715 + 537 1 252	34. 980 + 647 1 627	35. 882 + 619 1 501
36. 150 + 782 932	37. 833 + 978 1 811	38. 900 + 346 1 246	39. 260 + 768 1 028	40. 569 + 857 1 426	41. 719 + 238 957	42. 891 + 536 1 427

Find the sum

Complete all the activities (Addition).

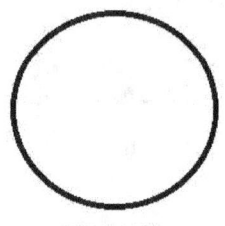

1. 729
  + 820
  ----

2. 775
  + 754
  ----

3. 852
  + 776
  ----

4. 160
  + 244
  ----

5. 912
  + 834
  ----

6. 171
  + 364
  ----

7. 987
  + 684
  ----

8. 361
  + 329
  ----

9. 986
  + 481
  ----

10. 348
   + 433
   ----

11. 723
   + 202
   ----

12. 845
   + 210
   ----

13. 452
   + 534
   ----

14. 358
   + 438
   ----

15. 482
   + 574
   ----

16. 555
   + 471
   ----

17. 221
   + 786
   ----

18. 451
   + 895
   ----

19. 761
   + 326
   ----

20. 794
   + 206
   ----

21. 559
   + 905
   ----

22. 869
   + 997
   ----

23. 907
   + 636
   ----

24. 359
   + 575
   ----

25. 507
   + 340
   ----

26. 871
   + 339
   ----

27. 890
   + 557
   ----

28. 812
   + 628
   ----

29. 974
   + 665
   ----

30. 402
   + 418
   ----

31. 518
   + 721
   ----

32. 761
   + 438
   ----

33. 380
   + 699
   ----

34. 243
   + 301
   ----

35. 332
   + 465
   ----

36. 567
   + 760
   ----

37. 686
   + 242
   ----

38. 874
   + 817
   ----

39. 332
   + 847
   ----

40. 303
   + 814
   ----

41. 276
   + 781
   ----

42. 930
   + 242

Find the sum

Complete all the activities (Addition).

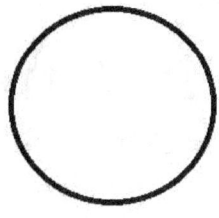

SCORE

1.	729	2.	775	3.	852	4.	160	5.	912	6.	171	7.	987
	+ 820		+ 754		+ 776		+ 244		+ 834		+ 364		+ 684
	1 549		1 529		1 628		404		1 746		535		1 671

8.	361	9.	986	10.	348	11.	723	12.	845	13.	452	14.	358
	+ 329		+ 481		+ 433		+ 202		+ 210		+ 534		+ 438
	690		1 467		781		925		1 055		986		796

15.	482	16.	555	17.	221	18.	451	19.	761	20.	794	21.	559
	+ 574		+ 471		+ 786		+ 895		+ 326		+ 206		+ 905
	1 056		1 026		1 007		1 346		1 087		1 000		1 464

22.	869	23.	907	24.	359	25.	507	26.	871	27.	890	28.	812
	+ 997		+ 636		+ 575		+ 340		+ 339		+ 557		+ 628
	1 866		1 543		934		847		1 210		1 447		1 440

29.	974	30.	402	31.	518	32.	761	33.	380	34.	243	35.	332
	+ 665		+ 418		+ 721		+ 438		+ 699		+ 301		+ 465
	1 639		820		1 239		1 199		1 079		544		797

36.	567	37.	686	38.	874	39.	332	40.	303	41.	276	42.	930
	+ 760		+ 242		+ 817		+ 847		+ 814		+ 781		+ 242
	1 327		928		1 691		1 179		1 117		1 057		1 172

Find the sum

Name:................................   Date:................................

Complete all the activities (Addition).

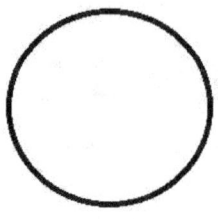

SCORE

1.   703
  + 316
	2. 580
	  + 663
		3. 973
		  + 823
			4. 920
			  + 197
				5. 923
				  + 476
					6. 687
					  + 837
						7. 244
						  + 531

8.   718
  + 308
	9. 493
	  + 158
		10. 343
		  + 142
			11. 772
			  + 211
				12. 651
				  + 748
					13. 683
					  + 994
						14. 396
						  + 227

15.   462
  + 587
	16. 243
	  + 736
		17. 790
		  + 619
			18. 734
			  + 163
				19. 219
				  + 580
					20. 327
					  + 509
						21. 544
						  + 470

22.   836
  + 310
	23. 553
	  + 137
		24. 673
		  + 800
			25. 229
			  + 994
				26. 421
				  + 217
					27. 539
					  + 888
						28. 469
						  + 592

29.   664
  + 832
	30. 979
	  + 978
		31. 527
		  + 793
			32. 733
			  + 436
				33. 333
				  + 950
					34. 147
					  + 624
						35. 796
						  + 654

36.   189
  + 912
	37. 831
	  + 833
		38. 596
		  + 402
			39. 374
			  + 441
				40. 294
				  + 646
					41. 421
					  + 595
						42. 997
						  + 467

Find the sum

Complete all the activities (Addition).

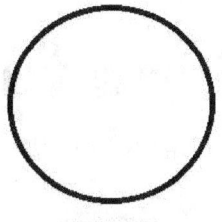

SCORE

1. 703 + 316 1 019	2. 580 + 663 1 243	3. 973 + 823 1 796	4. 920 + 197 1 117	5. 923 + 476 1 399	6. 687 + 837 1 524	7. 244 + 531 775
8. 718 + 308 1 026	9. 493 + 158 651	10. 343 + 142 485	11. 772 + 211 983	12. 651 + 748 1 399	13. 683 + 994 1 677	14. 396 + 227 623
15. 462 + 587 1 049	16. 243 + 736 979	17. 790 + 619 1 409	18. 734 + 163 897	19. 219 + 580 799	20. 327 + 509 836	21. 544 + 470 1 014
22. 836 + 310 1 146	23. 553 + 137 690	24. 673 + 800 1 473	25. 229 + 994 1 223	26. 421 + 217 638	27. 539 + 888 1 427	28. 469 + 592 1 061
29. 664 + 832 1 496	30. 979 + 978 1 957	31. 527 + 793 1 320	32. 733 + 436 1 169	33. 333 + 950 1 283	34. 147 + 624 771	35. 796 + 654 1 450
36. 189 + 912 1 101	37. 831 + 833 1 664	38. 596 + 402 998	39. 374 + 441 815	40. 294 + 646 940	41. 421 + 595 1 016	42. 997 + 467 1 464

Name:................................. 　　Date:...............................

Find the sum

Complete all the activities (Addition).

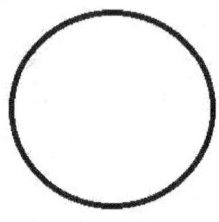

1.　266 + 125	2.　297 + 901	3.　250 + 242	4.　499 + 386	5.　600 + 347	6.　981 + 750	7.　736 + 500
8.　976 + 602	9.　989 + 149	10.　104 + 574	11.　905 + 508	12.　386 + 704	13.　392 + 167	14.　318 + 323
15.　637 + 433	16.　657 + 931	17.　795 + 366	18.　270 + 308	19.　753 + 245	20.　474 + 654	21.　256 + 801
22.　413 + 583	23.　851 + 845	24.　665 + 873	25.　464 + 691	26.　656 + 372	27.　591 + 523	28.　561 + 757
29.　739 + 592	30.　609 + 183	31.　923 + 276	32.　748 + 267	33.　689 + 150	34.　953 + 469	35.　330 + 769
36.　848 + 844	37.　674 + 449	38.　418 + 204	39.　195 + 801	40.　852 + 546	41.　501 + 444	42.　858 + 945

Find the sum

Complete all the activities (Addition).

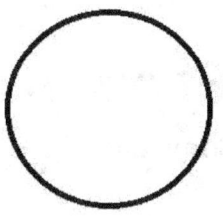

SCORE

1.	266	2.	297	3.	250	4.	499	5.	600	6.	981	7.	736
	+ 125		+ 901		+ 242		+ 386		+ 347		+ 750		+ 500
	391		1 198		492		885		947		1 731		1 236

8.	976	9.	989	10.	104	11.	905	12.	386	13.	392	14.	318
	+ 602		+ 149		+ 574		+ 508		+ 704		+ 167		+ 323
	1 578		1 138		678		1 413		1 090		559		641

15.	637	16.	657	17.	795	18.	270	19.	753	20.	474	21.	256
	+ 433		+ 931		+ 366		+ 308		+ 245		+ 654		+ 801
	1 070		1 588		1 161		578		998		1 128		1 057

22.	413	23.	851	24.	665	25.	464	26.	656	27.	591	28.	561
	+ 583		+ 845		+ 873		+ 691		+ 372		+ 523		+ 757
	996		1 696		1 538		1 155		1 028		1 114		1 318

29.	739	30.	609	31.	923	32.	748	33.	689	34.	953	35.	330
	+ 592		+ 183		+ 276		+ 267		+ 150		+ 469		+ 769
	1 331		792		1 199		1 015		839		1 422		1 099

36.	848	37.	674	38.	418	39.	195	40.	852	41.	501	42.	858
	+ 844		+ 449		+ 204		+ 801		+ 546		+ 444		+ 945
	1 692		1 123		622		996		1 398		945		1 803

Name:................................   Date:................................

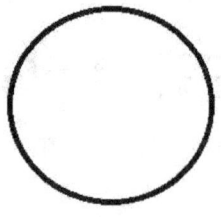

SCORE

Find the sum

Complete all the activities (Addition).

| 1. | 776
+ 385 | 2. | 440
+ 372 | 3. | 804
+ 512 | 4. | 568
+ 262 | 5. | 636
+ 391 | 6. | 729
+ 979 | 7. | 153
+ 245 |

| 8. | 893
+ 802 | 9. | 891
+ 296 | 10. | 643
+ 694 | 11. | 843
+ 811 | 12. | 697
+ 242 | 13. | 205
+ 457 | 14. | 661
+ 937 |

| 15. | 739
+ 163 | 16. | 681
+ 580 | 17. | 417
+ 803 | 18. | 490
+ 404 | 19. | 709
+ 681 | 20. | 664
+ 817 | 21. | 301
+ 686 |

| 22. | 308
+ 382 | 23. | 718
+ 438 | 24. | 150
+ 949 | 25. | 907
+ 150 | 26. | 654
+ 306 | 27. | 800
+ 264 | 28. | 202
+ 964 |

| 29. | 687
+ 297 | 30. | 267
+ 447 | 31. | 850
+ 630 | 32. | 522
+ 573 | 33. | 710
+ 250 | 34. | 662
+ 349 | 35. | 754
+ 589 |

| 36. | 193
+ 310 | 37. | 630
+ 825 | 38. | 493
+ 257 | 39. | 603
+ 415 | 40. | 523
+ 766 | 41. | 407
+ 712 | 42. | 129
+ 689 |

Find the sum

Complete all the activities (Addition).

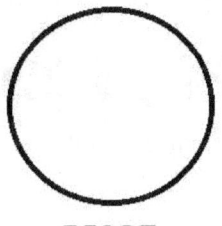

SCORE

1. 776 + 385 1 161	2. 440 + 372 812	3. 804 + 512 1 316	4. 568 + 262 830	5. 636 + 391 1 027	6. 729 + 979 1 708	7. 153 + 245 398
8. 893 + 802 1 695	9. 891 + 296 1 187	10. 643 + 694 1 337	11. 843 + 811 1 654	12. 697 + 242 939	13. 205 + 457 662	14. 661 + 937 1 598
15. 739 + 163 902	16. 681 + 580 1 261	17. 417 + 803 1 220	18. 490 + 404 894	19. 709 + 681 1 390	20. 664 + 817 1 481	21. 301 + 686 987
22. 308 + 382 690	23. 718 + 438 1 156	24. 150 + 949 1 099	25. 907 + 150 1 057	26. 654 + 306 960	27. 800 + 264 1 064	28. 202 + 964 1 166
29. 687 + 297 984	30. 267 + 447 714	31. 850 + 630 1 480	32. 522 + 573 1 095	33. 710 + 250 960	34. 662 + 349 1 011	35. 754 + 589 1 343
36. 193 + 310 503	37. 630 + 825 1 455	38. 493 + 257 750	39. 603 + 415 1 018	40. 523 + 766 1 289	41. 407 + 712 1 119	42. 129 + 689 818

Find the sum

Complete all the activities (Addition).

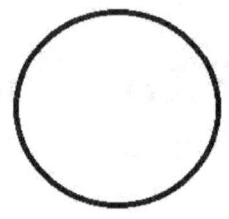

1.	508 + 475	2.	778 + 253	3.	701 + 526	4.	716 + 511	5.	515 + 318	6.	966 + 669	7.	265 + 735

8.	671 + 569	9.	775 + 800	10.	295 + 418	11.	956 + 444	12.	674 + 974	13.	770 + 944	14.	461 + 372

15.	773 + 499	16.	140 + 238	17.	601 + 832	18.	165 + 335	19.	470 + 510	20.	648 + 328	21.	873 + 680

22.	794 + 413	23.	858 + 823	24.	991 + 867	25.	743 + 953	26.	991 + 606	27.	517 + 711	28.	732 + 164

29.	800 + 803	30.	482 + 756	31.	165 + 444	32.	737 + 115	33.	278 + 827	34.	168 + 432	35.	670 + 437

36.	198 + 365	37.	252 + 966	38.	544 + 376	39.	649 + 362	40.	215 + 946	41.	952 + 593	42.	628 + 819

Find the sum

Complete all the activities (Addition).

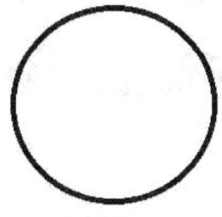

SCORE

| 1. | 508
+ 475
983 | 2. | 778
+ 253
1 031 | 3. | 701
+ 526
1 227 | 4. | 716
+ 511
1 227 | 5. | 515
+ 318
833 | 6. | 966
+ 669
1 635 | 7. | 265
+ 735
1 000 |

| 8. | 671
+ 569
1 240 | 9. | 775
+ 800
1 575 | 10. | 295
+ 418
713 | 11. | 956
+ 444
1 400 | 12. | 674
+ 974
1 648 | 13. | 770
+ 944
1 714 | 14. | 461
+ 372
833 |

| 15. | 773
+ 499
1 272 | 16. | 140
+ 238
378 | 17. | 601
+ 832
1 433 | 18. | 165
+ 335
500 | 19. | 470
+ 510
980 | 20. | 648
+ 328
976 | 21. | 873
+ 680
1 553 |

| 22. | 794
+ 413
1 207 | 23. | 858
+ 823
1 681 | 24. | 991
+ 867
1 858 | 25. | 743
+ 953
1 696 | 26. | 991
+ 606
1 597 | 27. | 517
+ 711
1 228 | 28. | 732
+ 164
896 |

| 29. | 800
+ 803
1 603 | 30. | 482
+ 756
1 238 | 31. | 165
+ 444
609 | 32. | 737
+ 115
852 | 33. | 278
+ 827
1 105 | 34. | 168
+ 432
600 | 35. | 670
+ 437
1 107 |

| 36. | 198
+ 365
563 | 37. | 252
+ 966
1 218 | 38. | 544
+ 376
920 | 39. | 649
+ 362
1 011 | 40. | 215
+ 946
1 161 | 41. | 952
+ 593
1 545 | 42. | 628
+ 819
1 447 |

Find the sum

Complete all the activities (Addition).

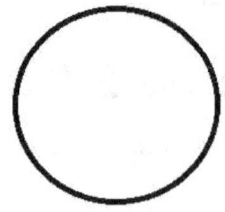

SCORE

1. 508 + 533	2. 721 + 889	3. 200 + 101	4. 534 + 666	5. 128 + 891	6. 264 + 529	7. 942 + 779

8. 235 + 131	9. 754 + 659	10. 297 + 406	11. 855 + 725	12. 454 + 550	13. 267 + 456	14. 514 + 538

15. 688 + 535	16. 802 + 570	17. 815 + 917	18. 607 + 744	19. 207 + 278	20. 331 + 670	21. 902 + 531

22. 337 + 360	23. 802 + 115	24. 432 + 869	25. 227 + 184	26. 292 + 393	27. 397 + 321	28. 963 + 781

29. 179 + 271	30. 723 + 743	31. 586 + 995	32. 977 + 592	33. 189 + 692	34. 712 + 125	35. 272 + 490

36. 924 + 576	37. 428 + 626	38. 504 + 424	39. 488 + 193	40. 747 + 251	41. 124 + 644	42. 412 + 604

Name:.................................   Date:.................................

Find the sum

Complete all the activities (Addition).

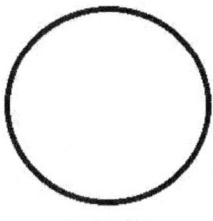

SCORE

1.	508	2.	721	3.	200	4.	534	5.	128	6.	264	7.	942
	+ 533		+ 889		+ 101		+ 666		+ 891		+ 529		+ 779
	1 041		1 610		301		1 200		1 019		793		1 721

8.	235	9.	754	10.	297	11.	855	12.	454	13.	267	14.	514
	+ 131		+ 659		+ 406		+ 725		+ 550		+ 456		+ 538
	366		1 413		703		1 580		1 004		723		1 052

15.	688	16.	802	17.	815	18.	607	19.	207	20.	331	21.	902
	+ 535		+ 570		+ 917		+ 744		+ 278		+ 670		+ 531
	1 223		1 372		1 732		1 351		485		1 001		1 433

22.	337	23.	802	24.	432	25.	227	26.	292	27.	397	28.	963
	+ 360		+ 115		+ 869		+ 184		+ 393		+ 321		+ 781
	697		917		1 301		411		685		718		1 744

29.	179	30.	723	31.	586	32.	977	33.	189	34.	712	35.	272
	+ 271		+ 743		+ 995		+ 592		+ 692		+ 125		+ 490
	450		1 466		1 581		1 569		881		837		762

36.	924	37.	428	38.	504	39.	488	40.	747	41.	124	42.	412
	+ 576		+ 626		+ 424		+ 193		+ 251		+ 644		+ 604
	1 500		1 054		928		681		998		768		1 016

Section 2

Name:...................................   Date:.....................................

Find the difference

Complete all the activities (Subtraction)

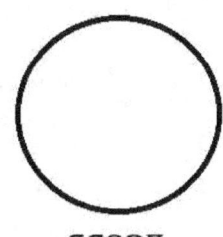

SCORE

1. 622 - 388 --------	2. 282 - 185 --------	3. 359 - 294 --------	4. 681 - 419 --------	5. 497 - 263 --------	6. 123 - 109 --------	7. 867 - 395 --------	8. 815 - 491 --------
9. 194 - 174 --------	10. 407 - 335 --------	11. 406 - 270 --------	12. 308 - 160 --------	13. 377 - 213 --------	14. 921 - 367 --------	15. 105 - 101 --------	16. 505 - 163 --------
17. 318 - 137 --------	18. 329 - 102 --------	19. 155 - 127 --------	20. 450 - 421 --------	21. 456 - 373 --------	22. 199 - 186 --------	23. 677 - 385 --------	24. 769 - 451 --------
25. 105 - 103 --------	26. 482 - 453 --------	27. 883 - 800 --------	28. 526 - 111 --------	29. 389 - 364 --------	30. 241 - 224 --------	31. 851 - 341 --------	32. 588 - 240 --------
33. 964 - 852 --------	34. 231 - 190 --------	35. 217 - 192 --------	36. 536 - 367 --------	37. 455 - 250 --------	38. 433 - 291 --------	39. 501 - 245 --------	40. 680 - 573 --------
41. 697 - 332 --------	42. 387 - 188 --------	43. 756 - 446 --------	44. 356 - 189 --------	45. 725 - 243 --------	46. 719 - 191 --------	47. 402 - 254 --------	48. 174 - 106 --------

Find the difference

Complete all the activities (Subtraction)

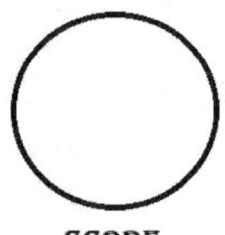

SCORE

1. 622 - 388 = 234	2. 282 - 185 = 97	3. 359 - 294 = 65	4. 681 - 419 = 262	5. 497 - 263 = 234	6. 123 - 109 = 14	7. 867 - 395 = 472	8. 815 - 491 = 324
9. 194 - 174 = 20	10. 407 - 335 = 72	11. 406 - 270 = 136	12. 308 - 160 = 148	13. 377 - 213 = 164	14. 921 - 367 = 554	15. 105 - 101 = 4	16. 505 - 163 = 342
17. 318 - 137 = 181	18. 329 - 102 = 227	19. 155 - 127 = 28	20. 450 - 421 = 29	21. 456 - 373 = 83	22. 199 - 186 = 13	23. 677 - 385 = 292	24. 769 - 451 = 318
25. 105 - 103 = 2	26. 482 - 453 = 29	27. 883 - 800 = 83	28. 526 - 111 = 415	29. 389 - 364 = 25	30. 241 - 224 = 17	31. 851 - 341 = 510	32. 588 - 240 = 348
33. 964 - 852 = 112	34. 231 - 190 = 41	35. 217 - 192 = 25	36. 536 - 367 = 169	37. 455 - 250 = 205	38. 433 - 291 = 142	39. 501 - 245 = 256	40. 680 - 573 = 107
41. 697 - 332 = 365	42. 387 - 188 = 199	43. 756 - 446 = 310	44. 356 - 189 = 167	45. 725 - 243 = 482	46. 719 - 191 = 528	47. 402 - 254 = 148	48. 174 - 106 = 68

Name:................................   Date:...............................

Find the difference

Complete all the activities (Subtraction)

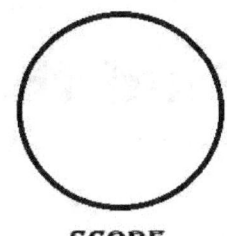

SCORE

1. 902 - 378 --------	2. 355 - 278 --------	3. 728 - 209 --------	4. 530 - 497 --------	5. 755 - 316 --------	6. 568 - 198 --------	7. 618 - 255 --------	8. 632 - 115 --------
9. 499 - 473 --------	10. 114 - 112 --------	11. 129 - 126 --------	12. 410 - 115 --------	13. 967 - 388 --------	14. 963 - 824 --------	15. 773 - 373 --------	16. 201 - 144 --------
17. 385 - 173 --------	18. 418 - 410 --------	19. 169 - 131 --------	20. 479 - 315 --------	21. 943 - 655 --------	22. 763 - 668 --------	23. 655 - 267 --------	24. 640 - 521 --------
25. 857 - 487 --------	26. 970 - 149 --------	27. 895 - 839 --------	28. 972 - 965 --------	29. 102 - 100 --------	30. 759 - 341 --------	31. 292 - 223 --------	32. 257 - 151 --------
33. 295 - 117 --------	34. 903 - 356 --------	35. 950 - 626 --------	36. 956 - 953 --------	37. 561 - 307 --------	38. 916 - 405 --------	39. 246 - 171 --------	40. 488 - 357 --------
41. 757 - 218 --------	42. 996 - 313 --------	43. 495 - 122 --------	44. 593 - 291 --------	45. 357 - 193 --------	46. 974 - 836 --------	47. 742 - 136 --------	48. 417 - 194 --------

Name:................................   Date:................................

Find the difference

Complete all the activities (Subtraction)

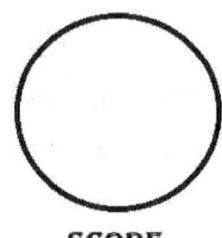

SCORE

| 1. | 902
- 378
524 | 2. | 355
- 278
77 | 3. | 728
- 209
519 | 4. | 530
- 497
33 | 5. | 755
- 316
439 | 6. | 568
- 198
370 | 7. | 618
- 255
363 | 8. | 632
- 115
517 |

| 9. | 499
- 473
26 | 10. | 114
- 112
2 | 11. | 129
- 126
3 | 12. | 410
- 115
295 | 13. | 967
- 388
579 | 14. | 963
- 824
139 | 15. | 773
- 373
400 | 16. | 201
- 144
57 |

| 17. | 385
- 173
212 | 18. | 418
- 410
8 | 19. | 169
- 131
38 | 20. | 479
- 315
164 | 21. | 943
- 655
288 | 22. | 763
- 668
95 | 23. | 655
- 267
388 | 24. | 640
- 521
119 |

| 25. | 857
- 487
370 | 26. | 970
- 149
821 | 27. | 895
- 839
56 | 28. | 972
- 965
7 | 29. | 102
- 100
2 | 30. | 759
- 341
418 | 31. | 292
- 223
69 | 32. | 257
- 151
106 |

| 33. | 295
- 117
178 | 34. | 903
- 356
547 | 35. | 950
- 626
324 | 36. | 956
- 953
3 | 37. | 561
- 307
254 | 38. | 916
- 405
511 | 39. | 246
- 171
75 | 40. | 488
- 357
131 |

| 41. | 757
- 218
539 | 42. | 996
- 313
683 | 43. | 495
- 122
373 | 44. | 593
- 291
302 | 45. | 357
- 193
164 | 46. | 974
- 836
138 | 47. | 742
- 136
606 | 48. | 417
- 194
223 |

Name:..................................   Date:..................................

Find the difference

Complete all the activities (Subtraction)

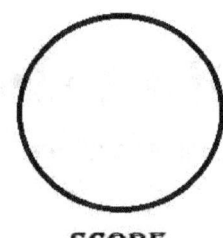

SCORE

1. 535 - 281 --------	2. 675 - 404 --------	3. 467 - 203 --------	4. 849 - 268 --------	5. 603 - 588 --------	6. 589 - 313 --------	7. 120 - 112 --------	8. 846 - 483 --------
9. 822 - 151 --------	10. 671 - 119 --------	11. 679 - 642 --------	12. 894 - 227 --------	13. 586 - 346 --------	14. 130 - 118 --------	15. 681 - 164 --------	16. 354 - 216 --------
17. 187 - 153 --------	18. 901 - 698 --------	19. 863 - 790 --------	20. 329 - 203 --------	21. 412 - 341 --------	22. 710 - 386 --------	23. 560 - 152 --------	24. 428 - 326 --------
25. 578 - 319 --------	26. 136 - 123 --------	27. 478 - 221 --------	28. 688 - 421 --------	29. 579 - 330 --------	30. 988 - 901 --------	31. 323 - 156 --------	32. 619 - 334 --------
33. 719 - 154 --------	34. 404 - 225 --------	35. 734 - 153 --------	36. 697 - 637 --------	37. 833 - 236 --------	38. 717 - 638 --------	39. 333 - 197 --------	40. 683 - 349 --------
41. 227 - 144 --------	42. 148 - 109 --------	43. 163 - 140 --------	44. 133 - 126 --------	45. 959 - 396 --------	46. 293 - 171 --------	47. 924 - 390 --------	48. 281 - 118 --------

Name:............................   Date:.................................

Find the difference

Complete all the activities (Subtraction)

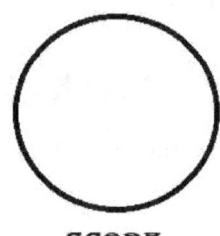

SCORE

1. 535 - 281 254	2. 675 - 404 271	3. 467 - 203 264	4. 849 - 268 581	5. 603 - 588 15	6. 589 - 313 276	7. 120 - 112 8	8. 846 - 483 363
9. 822 - 151 671	10. 671 - 119 552	11. 679 - 642 37	12. 894 - 227 667	13. 586 - 346 240	14. 130 - 118 12	15. 681 - 164 517	16. 354 - 216 138
17. 187 - 153 34	18. 901 - 698 203	19. 863 - 790 73	20. 329 - 203 126	21. 412 - 341 71	22. 710 - 386 324	23. 560 - 152 408	24. 428 - 326 102
25. 578 - 319 259	26. 136 - 123 13	27. 478 - 221 257	28. 688 - 421 267	29. 579 - 330 249	30. 988 - 901 87	31. 323 - 156 167	32. 619 - 334 285
33. 719 - 154 565	34. 404 - 225 179	35. 734 - 153 581	36. 697 - 637 60	37. 833 - 236 597	38. 717 - 638 79	39. 333 - 197 136	40. 683 - 349 334
41. 227 - 144 83	42. 148 - 109 39	43. 163 - 140 23	44. 133 - 126 7	45. 959 - 396 563	46. 293 - 171 122	47. 924 - 390 534	48. 281 - 118 163

Find the difference

Complete all the activities (Subtraction)

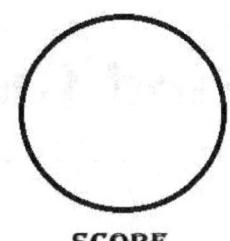

SCORE

| 1. 275
- 140
-------- | 2. 801
- 398
-------- | 3. 796
- 763
-------- | 4. 985
- 496
-------- | 5. 814
- 201
-------- | 6. 845
- 275
-------- | 7. 669
- 545
-------- | 8. 359
- 130
-------- |

| 9. 847
- 519
-------- | 10. 784
- 719
-------- | 11. 277
- 169
-------- | 12. 442
- 117
-------- | 13. 145
- 125
-------- | 14. 510
- 248
-------- | 15. 881
- 330
-------- | 16. 983
- 509
-------- |

| 17. 985
- 621
-------- | 18. 708
- 119
-------- | 19. 671
- 228
-------- | 20. 687
- 211
-------- | 21. 422
- 295
-------- | 22. 454
- 429
-------- | 23. 827
- 683
-------- | 24. 229
- 122
-------- |

| 25. 151
- 105
-------- | 26. 584
- 272
-------- | 27. 631
- 257
-------- | 28. 910
- 753
-------- | 29. 862
- 655
-------- | 30. 437
- 172
-------- | 31. 545
- 377
-------- | 32. 620
- 360
-------- |

| 33. 242
- 152
-------- | 34. 326
- 110
-------- | 35. 196
- 125
-------- | 36. 533
- 363
-------- | 37. 916
- 566
-------- | 38. 592
- 391
-------- | 39. 995
- 504
-------- | 40. 317
- 154
-------- |

| 41. 476
- 207
-------- | 42. 374
- 312
-------- | 43. 780
- 526
-------- | 44. 884
- 639
-------- | 45. 718
- 638
-------- | 46. 265
- 250
-------- | 47. 863
- 436
-------- | 48. 724
- 559
-------- |

Name:.................................. Date:..................................

Find the difference

Complete all the activities (Subtraction)

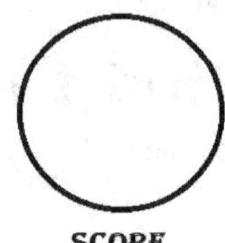

SCORE

1. 275 - 140 135	2. 801 - 398 403	3. 796 - 763 33	4. 985 - 496 489	5. 814 - 201 613	6. 845 - 275 570	7. 669 - 545 124	8. 359 - 130 229
9. 847 - 519 328	10. 784 - 719 65	11. 277 - 169 108	12. 442 - 117 325	13. 145 - 125 20	14. 510 - 248 262	15. 881 - 330 551	16. 983 - 509 474
17. 985 - 621 364	18. 708 - 119 589	19. 671 - 228 443	20. 687 - 211 476	21. 422 - 295 127	22. 454 - 429 25	23. 827 - 683 144	24. 229 - 122 107
25. 151 - 105 46	26. 584 - 272 312	27. 631 - 257 374	28. 910 - 753 157	29. 862 - 655 207	30. 437 - 172 265	31. 545 - 377 168	32. 620 - 360 260
33. 242 - 152 90	34. 326 - 110 216	35. 196 - 125 71	36. 533 - 363 170	37. 916 - 566 350	38. 592 - 391 201	39. 995 - 504 491	40. 317 - 154 163
41. 476 - 207 269	42. 374 - 312 62	43. 780 - 526 254	44. 884 - 639 245	45. 718 - 638 80	46. 265 - 250 15	47. 863 - 436 427	48. 724 - 559 165

Find the difference

Complete all the activities (Subtraction)

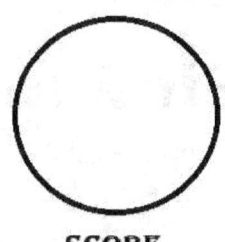

SCORE

Name:................................ Date:................................

1. 124 - 121	2. 793 - 407	3. 595 - 149	4. 160 - 148	5. 653 - 615	6. 503 - 108	7. 774 - 309	8. 829 - 389

9. 538 - 155	10. 814 - 251	11. 386 - 331	12. 768 - 455	13. 543 - 349	14. 420 - 176	15. 166 - 138	16. 781 - 419

17. 888 - 740	18. 608 - 151	19. 711 - 587	20. 728 - 300	21. 295 - 199	22. 241 - 216	23. 260 - 236	24. 944 - 419

25. 154 - 142	26. 477 - 373	27. 680 - 380	28. 475 - 183	29. 237 - 133	30. 954 - 943	31. 768 - 654	32. 752 - 126

33. 256 - 253	34. 228 - 220	35. 770 - 576	36. 745 - 545	37. 484 - 283	38. 994 - 826	39. 683 - 635	40. 417 - 303

41. 662 - 650	42. 259 - 146	43. 241 - 105	44. 839 - 392	45. 455 - 222	46. 953 - 476	47. 862 - 641	48. 617 - 300

Name:................................ Date:................................

Find the difference

Complete all the activities (Subtraction)

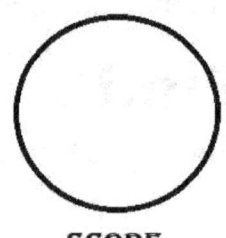

SCORE

1. 124 - 121 3	2. 793 - 407 386	3. 595 - 149 446	4. 160 - 148 12	5. 653 - 615 38	6. 503 - 108 395	7. 774 - 309 465	8. 829 - 389 440
9. 538 - 155 383	10. 814 - 251 563	11. 386 - 331 55	12. 768 - 455 313	13. 543 - 349 194	14. 420 - 176 244	15. 166 - 138 28	16. 781 - 419 362
17. 888 - 740 148	18. 608 - 151 457	19. 711 - 587 124	20. 728 - 300 428	21. 295 - 199 96	22. 241 - 216 25	23. 260 - 236 24	24. 944 - 419 525
25. 154 - 142 12	26. 477 - 373 104	27. 680 - 380 300	28. 475 - 183 292	29. 237 - 133 104	30. 954 - 943 11	31. 768 - 654 114	32. 752 - 126 626
33. 256 - 253 3	34. 228 - 220 8	35. 770 - 576 194	36. 745 - 545 200	37. 484 - 283 201	38. 994 - 826 168	39. 683 - 635 48	40. 417 - 303 114
41. 662 - 650 12	42. 259 - 146 113	43. 241 - 105 136	44. 839 - 392 447	45. 455 - 222 233	46. 953 - 476 477	47. 862 - 641 221	48. 617 - 300 317

Name:................................   Date:................................

Find the difference

Complete all the activities (Subtraction)

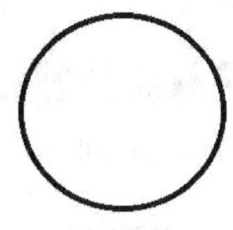

SCORE

1. 801 - 183 --------	2. 685 - 216 --------	3. 738 - 100 --------	4. 669 - 299 --------	5. 162 - 145 --------	6. 366 - 145 --------	7. 538 - 365 --------	8. 613 - 220 --------
9. 897 - 347 --------	10. 426 - 349 --------	11. 687 - 357 --------	12. 893 - 211 --------	13. 242 - 206 --------	14. 588 - 295 --------	15. 785 - 290 --------	16. 858 - 644 --------
17. 910 - 592 --------	18. 129 - 115 --------	19. 340 - 137 --------	20. 937 - 623 --------	21. 410 - 356 --------	22. 297 - 184 --------	23. 764 - 534 --------	24. 915 - 744 --------
25. 795 - 596 --------	26. 598 - 368 --------	27. 413 - 354 --------	28. 760 - 326 --------	29. 312 - 208 --------	30. 136 - 135 --------	31. 511 - 196 --------	32. 629 - 519 --------
33. 897 - 561 --------	34. 341 - 312 --------	35. 614 - 121 --------	36. 970 - 737 --------	37. 441 - 352 --------	38. 478 - 412 --------	39. 297 - 190 --------	40. 461 - 282 --------
41. 856 - 669 --------	42. 664 - 127 --------	43. 750 - 298 --------	44. 432 - 130 --------	45. 885 - 642 --------	46. 365 - 300 --------	47. 203 - 200 --------	48. 250 - 117 --------

Name:.............................. Date:..............................

Find the difference

Complete all the activities (Subtraction)

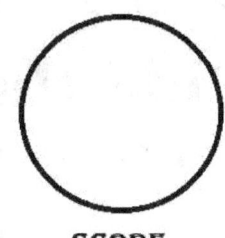

SCORE

1. 801 - 183 618	2. 685 - 216 469	3. 738 - 100 638	4. 669 - 299 370	5. 162 - 145 17	6. 366 - 145 221	7. 538 - 365 173	8. 613 - 220 393
9. 897 - 347 550	10. 426 - 349 77	11. 687 - 357 330	12. 893 - 211 682	13. 242 - 206 36	14. 588 - 295 293	15. 785 - 290 495	16. 858 - 644 214
17. 910 - 592 318	18. 129 - 115 14	19. 340 - 137 203	20. 937 - 623 314	21. 410 - 356 54	22. 297 - 184 113	23. 764 - 534 230	24. 915 - 744 171
25. 795 - 596 199	26. 598 - 368 230	27. 413 - 354 59	28. 760 - 326 434	29. 312 - 208 104	30. 136 - 135 1	31. 511 - 196 315	32. 629 - 519 110
33. 897 - 561 336	34. 341 - 312 29	35. 614 - 121 493	36. 970 - 737 233	37. 441 - 352 89	38. 478 - 412 66	39. 297 - 190 107	40. 461 - 282 179
41. 856 - 669 187	42. 664 - 127 537	43. 750 - 298 452	44. 432 - 130 302	45. 885 - 642 243	46. 365 - 300 65	47. 203 - 200 3	48. 250 - 117 133

Find the difference

Complete all the activities (Subtraction)

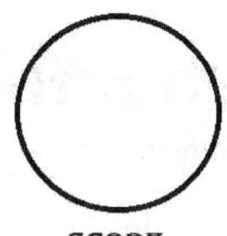

SCORE

1.	314 - 191 --------	2.	587 - 178 --------	3.	447 - 109 --------	4.	296 - 244 --------	5.	771 - 652 --------	6.	243 - 187 --------	7.	450 - 418 --------	8.	211 - 187 --------

9. 199   10. 208   11. 645   12. 162   13. 289   14. 106   15. 770   16. 654
   - 153   - 176   - 395   - 136   - 133   - 105   - 242   - 327
   --------   --------   --------   --------   --------   --------   --------   --------

17. 613   18. 629   19. 920   20. 207   21. 580   22. 371   23. 232   24. 731
   - 258   - 519   - 298   - 187   - 529   - 344   - 200   - 709
   --------   --------   --------   --------   --------   --------   --------   --------

25. 610   26. 680   27. 863   28. 555   29. 422   30. 954   31. 117   32. 988
   - 362   - 436   - 453   - 417   - 409   - 279   - 106   - 574
   --------   --------   --------   --------   --------   --------   --------   --------

33. 285   34. 135   35. 387   36. 881   37. 475   38. 253   39. 315   40. 490
   - 239   - 103   - 195   - 779   - 214   - 209   - 136   - 276
   --------   --------   --------   --------   --------   --------   --------   --------

41. 527   42. 382   43. 610   44. 604   45. 821   46. 490   47. 362   48. 815
   - 118   - 165   - 119   - 155   - 742   - 471   - 338   - 241
   --------   --------   --------   --------   --------   --------   --------   --------

Name:.............................. Date:................................

Find the difference

Complete all the activities (Subtraction)

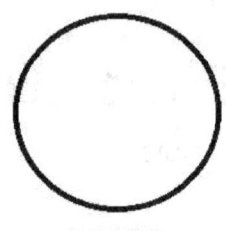

SCORE

1. 314	2. 587	3. 447	4. 296	5. 771	6. 243	7. 450	8. 211
- 191	- 178	- 109	- 244	- 652	- 187	- 418	- 187
123	409	338	52	119	56	32	24

9. 199	10. 208	11. 645	12. 162	13. 289	14. 106	15. 770	16. 654
- 153	- 176	- 395	- 136	- 133	- 105	- 242	- 327
46	32	250	26	156	1	528	327

17. 613	18. 629	19. 920	20. 207	21. 580	22. 371	23. 232	24. 731
- 258	- 519	- 298	- 187	- 529	- 344	- 200	- 709
355	110	622	20	51	27	32	22

25. 610	26. 680	27. 863	28. 555	29. 422	30. 954	31. 117	32. 988
- 362	- 436	- 453	- 417	- 409	- 279	- 106	- 574
248	244	410	138	13	675	11	414

33. 285	34. 135	35. 387	36. 881	37. 475	38. 253	39. 315	40. 490
- 239	- 103	- 195	- 779	- 214	- 209	- 136	- 276
46	32	192	102	261	44	179	214

41. 527	42. 382	43. 610	44. 604	45. 821	46. 490	47. 362	48. 815
- 118	- 165	- 119	- 155	- 742	- 471	- 338	- 241
409	217	491	449	79	19	24	574

Name:................................ Date:................................

Find the difference

Complete all the activities (Subtraction)

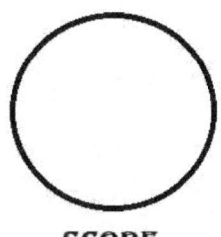

SCORE

| 1. 914
- 725
-------- | 2. 726
- 349
-------- | 3. 747
- 501
-------- | 4. 681
- 237
-------- | 5. 680
- 550
-------- | 6. 246
- 179
-------- | 7. 363
- 309
-------- | 8. 512
- 179
-------- |

| 9. 574
- 337
-------- | 10. 760
- 746
-------- | 11. 121
- 116
-------- | 12. 968
- 246
-------- | 13. 270
- 258
-------- | 14. 351
- 208
-------- | 15. 907
- 840
-------- | 16. 619
- 363
-------- |

| 17. 558
- 546
-------- | 18. 824
- 441
-------- | 19. 524
- 115
-------- | 20. 757
- 662
-------- | 21. 456
- 127
-------- | 22. 627
- 517
-------- | 23. 886
- 113
-------- | 24. 833
- 105
-------- |

| 25. 488
- 194
-------- | 26. 482
- 403
-------- | 27. 258
- 229
-------- | 28. 256
- 123
-------- | 29. 821
- 481
-------- | 30. 288
- 248
-------- | 31. 406
- 166
-------- | 32. 405
- 322
-------- |

| 33. 278
- 217
-------- | 34. 131
- 110
-------- | 35. 989
- 978
-------- | 36. 747
- 417
-------- | 37. 791
- 707
-------- | 38. 624
- 596
-------- | 39. 253
- 127
-------- | 40. 364
- 146
-------- |

| 41. 203
- 128
-------- | 42. 997
- 687
-------- | 43. 543
- 448
-------- | 44. 398
- 306
-------- | 45. 819
- 572
-------- | 46. 701
- 667
-------- | 47. 844
- 800
-------- | 48. 563
- 183
-------- |

Find the difference

Complete all the activities (Subtraction)

Name:................................   Date:................................

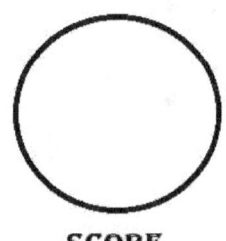

SCORE

| 1. 914
- 725
189 | 2. 726
- 349
377 | 3. 747
- 501
246 | 4. 681
- 237
444 | 5. 680
- 550
130 | 6. 246
- 179
67 | 7. 363
- 309
54 | 8. 512
- 179
333 |

| 9. 574
- 337
237 | 10. 760
- 746
14 | 11. 121
- 116
5 | 12. 968
- 246
722 | 13. 270
- 258
12 | 14. 351
- 208
143 | 15. 907
- 840
67 | 16. 619
- 363
256 |

| 17. 558
- 546
12 | 18. 824
- 441
383 | 19. 524
- 115
409 | 20. 757
- 662
95 | 21. 456
- 127
329 | 22. 627
- 517
110 | 23. 886
- 113
773 | 24. 833
- 105
728 |

| 25. 488
- 194
294 | 26. 482
- 403
79 | 27. 258
- 229
29 | 28. 256
- 123
133 | 29. 821
- 481
340 | 30. 288
- 248
40 | 31. 406
- 166
240 | 32. 405
- 322
83 |

| 33. 278
- 217
61 | 34. 131
- 110
21 | 35. 989
- 978
11 | 36. 747
- 417
330 | 37. 791
- 707
84 | 38. 624
- 596
28 | 39. 253
- 127
126 | 40. 364
- 146
218 |

| 41. 203
- 128
75 | 42. 997
- 687
310 | 43. 543
- 448
95 | 44. 398
- 306
92 | 45. 819
- 572
247 | 46. 701
- 667
34 | 47. 844
- 800
44 | 48. 563
- 183
380 |

Name:.................................. Date:....................................

Find the difference

Complete all the activities (Subtraction)

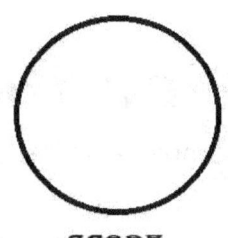

SCORE

1.   763
   - 669
   -------- 	
	2. 633
	   - 217
	   -------- 	
		3. 983
		   - 538
		   -------- 	
			4. 116
			   - 107
			   -------- 	
				5. 570
				   - 539
				   -------- 	
					6. 121
					   - 106
					   -------- 	
						7. 311
						   - 148
						   -------- 	
							8. 980
							   - 653
							   -------- 	

9.   311
   - 108
   -------- 	
	10. 707
	   - 141
	   -------- 	
		11. 297
		   - 235
		   -------- 	
			12. 312
			   - 188
			   -------- 	
				13. 391
				   - 200
				   -------- 	
					14. 567
					   - 255
					   -------- 	
						15. 387
						   - 323
						   -------- 	
							16. 696
							   - 593
							   -------- 	

17.   427
   - 232
   -------- 	
	18. 288
	   - 159
	   -------- 	
		19. 622
		   - 139
		   -------- 	
			20. 272
			   - 206
			   -------- 	
				21. 329
				   - 227
				   -------- 	
					22. 218
					   - 166
					   -------- 	
						23. 785
						   - 124
						   -------- 	
							24. 511
							   - 335
							   -------- 	

25.   710
   - 125
   -------- 	
	26. 737
	   - 604
	   -------- 	
		27. 339
		   - 207
		   -------- 	
			28. 963
			   - 145
			   -------- 	
				29. 660
				   - 201
				   -------- 	
					30. 906
					   - 615
					   -------- 	
						31. 978
						   - 702
						   -------- 	
							32. 213
							   - 184
							   -------- 	

33.   343
   - 147
   -------- 	
	34. 938
	   - 797
	   -------- 	
		35. 116
		   - 111
		   -------- 	
			36. 143
			   - 107
			   -------- 	
				37. 860
				   - 143
				   -------- 	
					38. 831
					   - 586
					   -------- 	
						39. 598
						   - 437
						   -------- 	
							40. 686
							   - 269
							   -------- 	

41.   896
   - 854
   -------- 	
	42. 688
	   - 263
	   -------- 	
		43. 161
		   - 153
		   -------- 	
			44. 605
			   - 563
			   -------- 	
				45. 499
				   - 162
				   -------- 	
					46. 720
					   - 348
					   -------- 	
						47. 260
						   - 188
						   -------- 	
							48. 677
							   - 428
							   -------- 	

Find the difference

Complete all the activities (Subtraction)

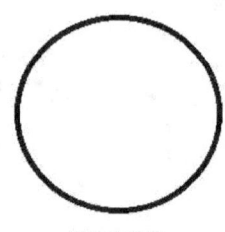

SCORE

| 1. 763
- 669
94 | 2. 633
- 217
416 | 3. 983
- 538
445 | 4. 116
- 107
9 | 5. 570
- 539
31 | 6. 121
- 106
15 | 7. 311
- 148
163 | 8. 980
- 653
327 |

| 9. 311
- 108
203 | 10. 707
- 141
566 | 11. 297
- 235
62 | 12. 312
- 188
124 | 13. 391
- 200
191 | 14. 567
- 255
312 | 15. 387
- 323
64 | 16. 696
- 593
103 |

| 17. 427
- 232
195 | 18. 288
- 159
129 | 19. 622
- 139
483 | 20. 272
- 206
66 | 21. 329
- 227
102 | 22. 218
- 166
52 | 23. 785
- 124
661 | 24. 511
- 335
176 |

| 25. 710
- 125
585 | 26. 737
- 604
133 | 27. 339
- 207
132 | 28. 963
- 145
818 | 29. 660
- 201
459 | 30. 906
- 615
291 | 31. 978
- 702
276 | 32. 213
- 184
29 |

| 33. 343
- 147
196 | 34. 938
- 797
141 | 35. 116
- 111
5 | 36. 143
- 107
36 | 37. 860
- 143
717 | 38. 831
- 586
245 | 39. 598
- 437
161 | 40. 686
- 269
417 |

| 41. 896
- 854
42 | 42. 688
- 263
425 | 43. 161
- 153
8 | 44. 605
- 563
42 | 45. 499
- 162
337 | 46. 720
- 348
372 | 47. 260
- 188
72 | 48. 677
- 428
249 |

Name:.. Date:..

Find the difference

Complete all the activities (Subtraction)

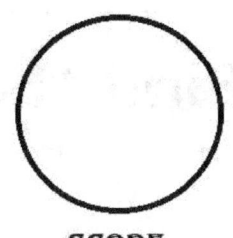

SCORE

| 1. | 485
- 248
-------- | 2. | 358
- 246
-------- | 3. | 192
- 158
-------- | 4. | 598
- 106
-------- | 5. | 993
- 589
-------- | 6. | 950
- 278
-------- | 7. | 670
- 268
-------- | 8. | 548
- 523
-------- |

| 9. | 575
- 389
-------- | 10. | 778
- 120
-------- | 11. | 129
- 125
-------- | 12. | 711
- 343
-------- | 13. | 669
- 109
-------- | 14. | 449
- 401
-------- | 15. | 977
- 668
-------- | 16. | 984
- 360
-------- |

| 17. | 473
- 192
-------- | 18. | 313
- 136
-------- | 19. | 451
- 130
-------- | 20. | 267
- 143
-------- | 21. | 453
- 327
-------- | 22. | 356
- 233
-------- | 23. | 934
- 394
-------- | 24. | 355
- 138
-------- |

| 25. | 691
- 541
-------- | 26. | 884
- 546
-------- | 27. | 816
- 539
-------- | 28. | 582
- 380
-------- | 29. | 734
- 664
-------- | 30. | 939
- 173
-------- | 31. | 314
- 248
-------- | 32. | 132
- 111
-------- |

| 33. | 608
- 431
-------- | 34. | 815
- 378
-------- | 35. | 153
- 117
-------- | 36. | 599
- 322
-------- | 37. | 480
- 288
-------- | 38. | 672
- 323
-------- | 39. | 605
- 181
-------- | 40. | 523
- 219
-------- |

| 41. | 758
- 258
-------- | 42. | 284
- 142
-------- | 43. | 797
- 388
-------- | 44. | 547
- 458
-------- | 45. | 738
- 518
-------- | 46. | 517
- 309
-------- | 47. | 236
- 113
-------- | 48. | 119
- 113
-------- |

Name:.................................   Date:...............................

Find the difference

Complete all the activities (Subtraction)

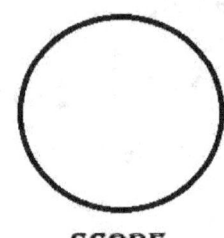

SCORE

1. 485	2. 358	3. 192	4. 598	5. 993	6. 950	7. 670	8. 548
- 248	- 246	- 158	- 106	- 589	- 278	- 268	- 523
237	112	34	492	404	672	402	25

9. 575	10. 778	11. 129	12. 711	13. 669	14. 449	15. 977	16. 984
- 389	- 120	- 125	- 343	- 109	- 401	- 668	- 360
186	658	4	368	560	48	309	624

17. 473	18. 313	19. 451	20. 267	21. 453	22. 356	23. 934	24. 355
- 192	- 136	- 130	- 143	- 327	- 233	- 394	- 138
281	177	321	124	126	123	540	217

25. 691	26. 884	27. 816	28. 582	29. 734	30. 939	31. 314	32. 132
- 541	- 546	- 539	- 380	- 664	- 173	- 248	- 111
150	338	277	202	70	766	66	21

33. 608	34. 815	35. 153	36. 599	37. 480	38. 672	39. 605	40. 523
- 431	- 378	- 117	- 322	- 288	- 323	- 181	- 219
177	437	36	277	192	349	424	304

41. 758	42. 284	43. 797	44. 547	45. 738	46. 517	47. 236	48. 119
- 258	- 142	- 388	- 458	- 518	- 309	- 113	- 113
500	142	409	89	220	208	123	6

Find the difference

Complete all the activities (Subtraction)

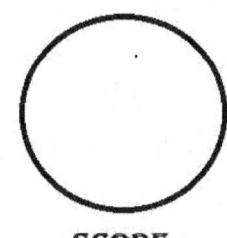

1.	557 - 123	2.	713 - 478	3.	144 - 107	4.	362 - 117	5.	326 - 112	6.	237 - 197	7.	997 - 625	8.	588 - 573
9.	167 - 121	10.	105 - 103	11.	479 - 374	12.	267 - 146	13.	597 - 244	14.	304 - 256	15.	683 - 294	16.	121 - 103
17.	280 - 234	18.	279 - 249	19.	859 - 327	20.	330 - 277	21.	524 - 104	22.	264 - 221	23.	357 - 343	24.	326 - 310
25.	389 - 174	26.	103 - 101	27.	371 - 213	28.	908 - 735	29.	877 - 418	30.	157 - 143	31.	948 - 548	32.	475 - 444
33.	127 - 127	34.	558 - 217	35.	448 - 347	36.	212 - 187	37.	156 - 113	38.	196 - 140	39.	585 - 317	40.	988 - 639
41.	135 - 110	42.	497 - 256	43.	397 - 273	44.	868 - 702	45.	859 - 413	46.	970 - 479	47.	312 - 205	48.	844 - 802

Find the difference

Complete all the activities (Subtraction)

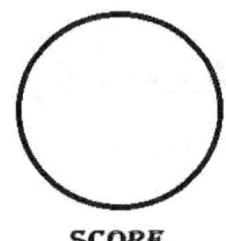

SCORE

1. 557 - 123 434	2. 713 - 478 235	3. 144 - 107 37	4. 362 - 117 245	5. 326 - 112 214	6. 237 - 197 40	7. 997 - 625 372	8. 588 - 573 15
9. 167 - 121 46	10. 105 - 103 2	11. 479 - 374 105	12. 267 - 146 121	13. 597 - 244 353	14. 304 - 256 48	15. 683 - 294 389	16. 121 - 103 18
17. 280 - 234 46	18. 279 - 249 30	19. 859 - 327 532	20. 330 - 277 53	21. 524 - 104 420	22. 264 - 221 43	23. 357 - 343 14	24. 326 - 310 16
25. 389 - 174 215	26. 103 - 101 2	27. 371 - 213 158	28. 908 - 735 173	29. 877 - 418 459	30. 157 - 143 14	31. 948 - 548 400	32. 475 - 444 31
33. 127 - 127 0	34. 558 - 217 341	35. 448 - 347 101	36. 212 - 187 25	37. 156 - 113 43	38. 196 - 140 56	39. 585 - 317 268	40. 988 - 639 349
41. 135 - 110 25	42. 497 - 256 241	43. 397 - 273 124	44. 868 - 702 166	45. 859 - 413 446	46. 970 - 479 491	47. 312 - 205 107	48. 844 - 802 42

Name:..   Date:...................................

Find the difference

Complete all the activities (Subtraction)

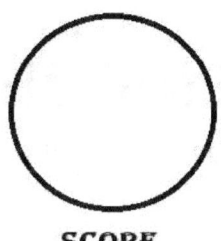

SCORE

1.   632
   - 241
   -------- 	
	2. 737
	   - 284
	   -------- 	
		3. 493
		   - 491
		   -------- 	
			4. 587
			   - 398
			   -------- 	
				5. 217
				   - 127
				   -------- 	
					6. 439
					   - 122
					   -------- 	
						7. 379
						   - 236
						   -------- 	
							8. 321
							   - 192
							   -------- 	

9.   478
   - 273
   -------- 	
	10. 748
	   - 518
	   -------- 	
		11. 178
		   - 102
		   -------- 	
			12. 629
			   - 336
			   -------- 	
				13. 228
				   - 193
				   -------- 	
					14. 469
					   - 460
					   -------- 	
						15. 768
						   - 560
						   -------- 	
							16. 586
							   - 375
							   -------- 	

17.   467
   - 160
   -------- 	
	18. 838
	   - 303
	   -------- 	
		19. 764
		   - 564
		   -------- 	
			20. 252
			   - 106
			   -------- 	
				21. 779
				   - 664
				   -------- 	
					22. 337
					   - 124
					   -------- 	
						23. 286
						   - 247
						   -------- 	
							24. 526
							   - 253
							   -------- 	

25.   250
   - 182
   -------- 	
	26. 114
	   - 111
	   -------- 	
		27. 797
		   - 227
		   -------- 	
			28. 870
			   - 538
			   -------- 	
				29. 709
				   - 259
				   -------- 	
					30. 232
					   - 194
					   -------- 	
						31. 626
						   - 399
						   -------- 	
							32. 880
							   - 444
							   -------- 	

33.   718
   - 601
   -------- 	
	34. 692
	   - 187
	   -------- 	
		35. 698
		   - 681
		   -------- 	
			36. 711
			   - 541
			   -------- 	
				37. 359
				   - 276
				   -------- 	
					38. 900
					   - 305
					   -------- 	
						39. 782
						   - 431
						   -------- 	
							40. 316
							   - 176
							   -------- 	

41.   327
   - 112
   -------- 	
	42. 826
	   - 241
	   -------- 	
		43. 706
		   - 562
		   -------- 	
			44. 811
			   - 261
			   -------- 	
				45. 376
				   - 170
				   -------- 	
					46. 637
					   - 356
					   -------- 	
						47. 597
						   - 327
						   -------- 	
							48. 511
							   - 177
							   -------- 	

Name:.............................   Date:.............................

Find the difference

Complete all the activities (Subtraction)

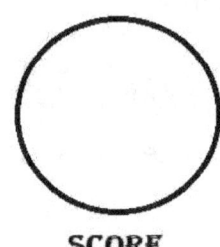

SCORE

1. 632 − 241 = 391	2. 737 − 284 = 453	3. 493 − 491 = 2	4. 587 − 398 = 189	5. 217 − 127 = 90	6. 439 − 122 = 317	7. 379 − 236 = 143	8. 321 − 192 = 129
9. 478 − 273 = 205	10. 748 − 518 = 230	11. 178 − 102 = 76	12. 629 − 336 = 293	13. 228 − 193 = 35	14. 469 − 460 = 9	15. 768 − 560 = 208	16. 586 − 375 = 211
17. 467 − 160 = 307	18. 838 − 303 = 535	19. 764 − 564 = 200	20. 252 − 106 = 146	21. 779 − 664 = 115	22. 337 − 124 = 213	23. 286 − 247 = 39	24. 526 − 253 = 273
25. 250 − 182 = 68	26. 114 − 111 = 3	27. 797 − 227 = 570	28. 870 − 538 = 332	29. 709 − 259 = 450	30. 232 − 194 = 38	31. 626 − 399 = 227	32. 880 − 444 = 436
33. 718 − 601 = 117	34. 692 − 187 = 505	35. 698 − 681 = 17	36. 711 − 541 = 170	37. 359 − 276 = 83	38. 900 − 305 = 595	39. 782 − 431 = 351	40. 316 − 176 = 140
41. 327 − 112 = 215	42. 826 − 241 = 585	43. 706 − 562 = 144	44. 811 − 261 = 550	45. 376 − 170 = 206	46. 637 − 356 = 281	47. 597 − 327 = 270	48. 511 − 177 = 334

Find the difference

Complete all the activities (Subtraction)

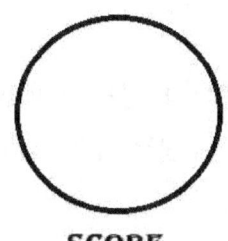

SCORE

| 1. 966
- 813
-------- | 2. 473
- 248
-------- | 3. 217
- 138
-------- | 4. 723
- 221
-------- | 5. 882
- 330
-------- | 6. 540
- 142
-------- | 7. 362
- 257
-------- | 8. 828
- 239
-------- |

| 9. 119
- 115
-------- | 10. 514
- 320
-------- | 11. 783
- 560
-------- | 12. 789
- 737
-------- | 13. 118
- 108
-------- | 14. 854
- 411
-------- | 15. 436
- 168
-------- | 16. 251
- 226
-------- |

| 17. 299
- 240
-------- | 18. 364
- 347
-------- | 19. 164
- 154
-------- | 20. 302
- 287
-------- | 21. 424
- 299
-------- | 22. 392
- 211
-------- | 23. 345
- 270
-------- | 24. 940
- 694
-------- |

| 25. 498
- 262
-------- | 26. 509
- 411
-------- | 27. 996
- 929
-------- | 28. 485
- 211
-------- | 29. 151
- 125
-------- | 30. 438
- 336
-------- | 31. 539
- 223
-------- | 32. 837
- 592
-------- |

| 33. 213
- 174
-------- | 34. 887
- 564
-------- | 35. 376
- 341
-------- | 36. 208
- 119
-------- | 37. 611
- 373
-------- | 38. 255
- 123
-------- | 39. 414
- 221
-------- | 40. 686
- 291
-------- |

| 41. 547
- 159
-------- | 42. 972
- 944
-------- | 43. 772
- 374
-------- | 44. 851
- 323
-------- | 45. 346
- 101
-------- | 46. 890
- 559
-------- | 47. 522
- 322
-------- | 48. 493
- 131
-------- |

Find the difference

Complete all the activities (Subtraction)

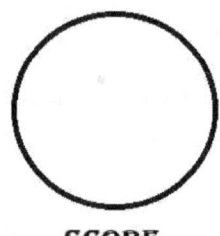

SCORE

1. 966	2. 473	3. 217	4. 723	5. 882	6. 540	7. 362	8. 828
- 813	- 248	- 138	- 221	- 330	- 142	- 257	- 239
153	225	79	502	552	398	105	589

9. 119	10. 514	11. 783	12. 789	13. 118	14. 854	15. 436	16. 251
- 115	- 320	- 560	- 737	- 108	- 411	- 168	- 226
4	194	223	52	10	443	268	25

17. 299	18. 364	19. 164	20. 302	21. 424	22. 392	23. 345	24. 940
- 240	- 347	- 154	- 287	- 299	- 211	- 270	- 694
59	17	10	15	125	181	75	246

25. 498	26. 509	27. 996	28. 485	29. 151	30. 438	31. 539	32. 837
- 262	- 411	- 929	- 211	- 125	- 336	- 223	- 592
236	98	67	274	26	102	316	245

33. 213	34. 887	35. 376	36. 208	37. 611	38. 255	39. 414	40. 686
- 174	- 564	- 341	- 119	- 373	- 123	- 221	- 291
39	323	35	89	238	132	193	395

41. 547	42. 972	43. 772	44. 851	45. 346	46. 890	47. 522	48. 493
- 159	- 944	- 374	- 323	- 101	- 559	- 322	- 131
388	28	398	528	245	331	200	362

Name:............................... Date:...............................

Find the difference

Complete all the activities (Subtraction)

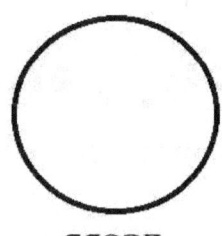

SCORE

| 1. | 813
- 569
-------- | 2. | 328
- 240
-------- | 3. | 837
- 354
-------- | 4. | 460
- 305
-------- | 5. | 357
- 172
-------- | 6. | 273
- 180
-------- | 7. | 777
- 291
-------- | 8. | 945
- 666
-------- |

| 9. | 853
- 138
-------- | 10. | 279
- 164
-------- | 11. | 379
- 338
-------- | 12. | 927
- 694
-------- | 13. | 840
- 528
-------- | 14. | 385
- 186
-------- | 15. | 995
- 583
-------- | 16. | 736
- 126
-------- |

| 17. | 188
- 108
-------- | 18. | 418
- 320
-------- | 19. | 219
- 104
-------- | 20. | 555
- 170
-------- | 21. | 758
- 364
-------- | 22. | 713
- 387
-------- | 23. | 294
- 172
-------- | 24. | 589
- 585
-------- |

| 25. | 157
- 112
-------- | 26. | 768
- 380
-------- | 27. | 394
- 124
-------- | 28. | 791
- 750
-------- | 29. | 763
- 455
-------- | 30. | 812
- 271
-------- | 31. | 125
- 102
-------- | 32. | 615
- 484
-------- |

| 33. | 614
- 178
-------- | 34. | 427
- 316
-------- | 35. | 935
- 307
-------- | 36. | 583
- 487
-------- | 37. | 589
- 123
-------- | 38. | 660
- 609
-------- | 39. | 253
- 169
-------- | 40. | 552
- 303
-------- |

| 41. | 131
- 110
-------- | 42. | 315
- 269
-------- | 43. | 816
- 306
-------- | 44. | 741
- 110
-------- | 45. | 391
- 220
-------- | 46. | 958
- 612
-------- | 47. | 320
- 158
-------- | 48. | 213
- 172
-------- |

Find the difference

Complete all the activities (Subtraction)

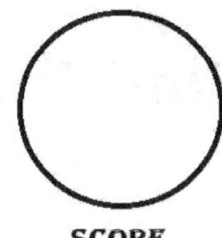

SCORE

1. 813	2. 328	3. 837	4. 460	5. 357	6. 273	7. 777	8. 945
- 569	- 240	- 354	- 305	- 172	- 180	- 291	- 666
244	88	483	155	185	93	486	279

9. 853	10. 279	11. 379	12. 927	13. 840	14. 385	15. 995	16. 736
- 138	- 164	- 338	- 694	- 528	- 186	- 583	- 126
715	115	41	233	312	199	412	610

17. 188	18. 418	19. 219	20. 555	21. 758	22. 713	23. 294	24. 589
- 108	- 320	- 104	- 170	- 364	- 387	- 172	- 585
80	98	115	385	394	326	122	4

25. 157	26. 768	27. 394	28. 791	29. 763	30. 812	31. 125	32. 615
- 112	- 380	- 124	- 750	- 455	- 271	- 102	- 484
45	388	270	41	308	541	23	131

33. 614	34. 427	35. 935	36. 583	37. 589	38. 660	39. 253	40. 552
- 178	- 316	- 307	- 487	- 123	- 609	- 169	- 303
436	111	628	96	466	51	84	249

41. 131	42. 315	43. 816	44. 741	45. 391	46. 958	47. 320	48. 213
- 110	- 269	- 306	- 110	- 220	- 612	- 158	- 172
21	46	510	631	171	346	162	41

Find the difference

Complete all the activities (Subtraction)

Name:.............................. Date:................................

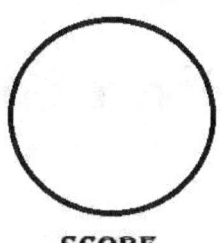

SCORE

1. 844 - 796	2. 907 - 151	3. 109 - 108	4. 156 - 146	5. 704 - 669	6. 887 - 321	7. 512 - 454	8. 793 - 299
9. 130 - 110	10. 285 - 230	11. 871 - 706	12. 954 - 342	13. 494 - 237	14. 402 - 256	15. 364 - 348	16. 600 - 239
17. 928 - 918	18. 174 - 117	19. 598 - 319	20. 495 - 160	21. 336 - 200	22. 204 - 189	23. 471 - 260	24. 602 - 333
25. 676 - 484	26. 607 - 557	27. 154 - 123	28. 800 - 129	29. 986 - 644	30. 662 - 472	31. 593 - 369	32. 568 - 442
33. 446 - 307	34. 424 - 252	35. 798 - 475	36. 759 - 491	37. 912 - 144	38. 495 - 271	39. 704 - 158	40. 883 - 105
41. 810 - 718	42. 887 - 593	43. 439 - 322	44. 296 - 198	45. 289 - 108	46. 323 - 310	47. 832 - 670	48. 641 - 488

Name:............................. Date:...................................

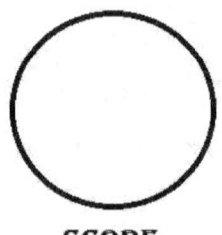

SCORE

Find the difference

Complete all the activities (Subtraction)

1. 844 - 796 48	2. 907 - 151 756	3. 109 - 108 1	4. 156 - 146 10	5. 704 - 669 35	6. 887 - 321 566	7. 512 - 454 58	8. 793 - 299 494
9. 130 - 110 20	10. 285 - 230 55	11. 871 - 706 165	12. 954 - 342 612	13. 494 - 237 257	14. 402 - 256 146	15. 364 - 348 16	16. 600 - 239 361
17. 928 - 918 10	18. 174 - 117 57	19. 598 - 319 279	20. 495 - 160 335	21. 336 - 200 136	22. 204 - 189 15	23. 471 - 260 211	24. 602 - 333 269
25. 676 - 484 192	26. 607 - 557 50	27. 154 - 123 31	28. 800 - 129 671	29. 986 - 644 342	30. 662 - 472 190	31. 593 - 369 224	32. 568 - 442 126
33. 446 - 307 139	34. 424 - 252 172	35. 798 - 475 323	36. 759 - 491 268	37. 912 - 144 768	38. 495 - 271 224	39. 704 - 158 546	40. 883 - 105 778
41. 810 - 718 92	42. 887 - 593 294	43. 439 - 322 117	44. 296 - 198 98	45. 289 - 108 181	46. 323 - 310 13	47. 832 - 670 162	48. 641 - 488 153

Name:.................................   Date:.................................

Find the difference

Complete all the activities (Subtraction)

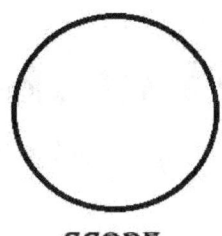

SCORE

1. 912 - 412 -------	2. 521 - 443 -------	3. 165 - 106 -------	4. 719 - 165 -------	5. 940 - 915 -------	6. 997 - 105 -------	7. 984 - 406 -------	8. 432 - 395 -------
9. 768 - 513 -------	10. 884 - 632 -------	11. 703 - 639 -------	12. 419 - 333 -------	13. 441 - 394 -------	14. 334 - 272 -------	15. 359 - 305 -------	16. 800 - 188 -------
17. 507 - 394 -------	18. 576 - 237 -------	19. 664 - 131 -------	20. 941 - 690 -------	21. 264 - 226 -------	22. 722 - 353 -------	23. 691 - 226 -------	24. 563 - 500 -------
25. 704 - 317 -------	26. 277 - 135 -------	27. 914 - 852 -------	28. 142 - 100 -------	29. 127 - 120 -------	30. 492 - 182 -------	31. 844 - 809 -------	32. 834 - 203 -------
33. 271 - 249 -------	34. 597 - 230 -------	35. 834 - 160 -------	36. 336 - 262 -------	37. 714 - 205 -------	38. 545 - 500 -------	39. 282 - 272 -------	40. 166 - 111 -------
41. 412 - 224 -------	42. 157 - 153 -------	43. 636 - 412 -------	44. 297 - 204 -------	45. 707 - 416 -------	46. 331 - 305 -------	47. 657 - 504 -------	48. 886 - 281 -------

Name:..................................   Date:..................................

Find the difference

Complete all the activities (Subtraction)

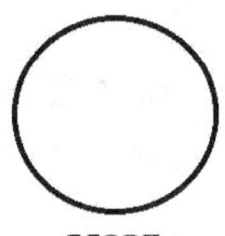

SCORE

1. 912 - 412 500	2. 521 - 443 78	3. 165 - 106 59	4. 719 - 165 554	5. 940 - 915 25	6. 997 - 105 892	7. 984 - 406 578	8. 432 - 395 37
9. 768 - 513 255	10. 884 - 632 252	11. 703 - 639 64	12. 419 - 333 86	13. 441 - 394 47	14. 334 - 272 62	15. 359 - 305 54	16. 800 - 188 612
17. 507 - 394 113	18. 576 - 237 339	19. 664 - 131 533	20. 941 - 690 251	21. 264 - 226 38	22. 722 - 353 369	23. 691 - 226 465	24. 563 - 500 63
25. 704 - 317 387	26. 277 - 135 142	27. 914 - 852 62	28. 142 - 100 42	29. 127 - 120 7	30. 492 - 182 310	31. 844 - 809 35	32. 834 - 203 631
33. 271 - 249 22	34. 597 - 230 367	35. 834 - 160 674	36. 336 - 262 74	37. 714 - 205 509	38. 545 - 500 45	39. 282 - 272 10	40. 166 - 111 55
41. 412 - 224 188	42. 157 - 153 4	43. 636 - 412 224	44. 297 - 204 93	45. 707 - 416 291	46. 331 - 305 26	47. 657 - 504 153	48. 886 - 281 605

Find the difference

Complete all the activities (Subtraction)

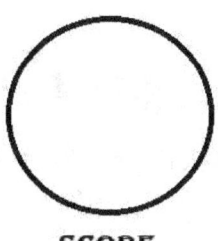

SCORE

| 1. | 959
- 574
-------- | 2. | 295
- 253
-------- | 3. | 693
- 250
-------- | 4. | 143
- 125
-------- | 5. | 186
- 144
-------- | 6. | 562
- 440
-------- | 7. | 485
- 157
-------- | 8. | 826
- 160
-------- |

| 9. | 698
- 261
-------- | 10. | 875
- 434
-------- | 11. | 899
- 808
-------- | 12. | 589
- 398
-------- | 13. | 688
- 114
-------- | 14. | 627
- 256
-------- | 15. | 842
- 378
-------- | 16. | 161
- 158
-------- |

| 17. | 367
- 171
-------- | 18. | 949
- 652
-------- | 19. | 757
- 290
-------- | 20. | 556
- 143
-------- | 21. | 413
- 249
-------- | 22. | 849
- 165
-------- | 23. | 161
- 130
-------- | 24. | 190
- 147
-------- |

| 25. | 741
- 212
-------- | 26. | 300
- 170
-------- | 27. | 876
- 443
-------- | 28. | 407
- 121
-------- | 29. | 106
- 104
-------- | 30. | 217
- 174
-------- | 31. | 699
- 550
-------- | 32. | 787
- 596
-------- |

| 33. | 497
- 321
-------- | 34. | 494
- 164
-------- | 35. | 283
- 129
-------- | 36. | 416
- 219
-------- | 37. | 785
- 626
-------- | 38. | 175
- 131
-------- | 39. | 147
- 121
-------- | 40. | 586
- 182
-------- |

| 41. | 931
- 149
-------- | 42. | 624
- 441
-------- | 43. | 236
- 198
-------- | 44. | 185
- 173
-------- | 45. | 931
- 827
-------- | 46. | 679
- 171
-------- | 47. | 243
- 225
-------- | 48. | 485
- 330
-------- |

Name:..............................   Date:..............................

Find the difference

Complete all the activities (Subtraction)

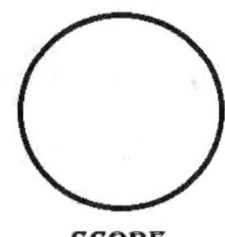

SCORE

1. 959 - 574 385	2. 295 - 253 42	3. 693 - 250 443	4. 143 - 125 18	5. 186 - 144 42	6. 562 - 440 122	7. 485 - 157 328	8. 826 - 160 666

9. 698 - 261 437	10. 875 - 434 441	11. 899 - 808 91	12. 589 - 398 191	13. 688 - 114 574	14. 627 - 256 371	15. 842 - 378 464	16. 161 - 158 3

17. 367 - 171 196	18. 949 - 652 297	19. 757 - 290 467	20. 556 - 143 413	21. 413 - 249 164	22. 849 - 165 684	23. 161 - 130 31	24. 190 - 147 43

25. 741 - 212 529	26. 300 - 170 130	27. 876 - 443 433	28. 407 - 121 286	29. 106 - 104 2	30. 217 - 174 43	31. 699 - 550 149	32. 787 - 596 191

33. 497 - 321 176	34. 494 - 164 330	35. 283 - 129 154	36. 416 - 219 197	37. 785 - 626 159	38. 175 - 131 44	39. 147 - 121 26	40. 586 - 182 404

41. 931 - 149 782	42. 624 - 441 183	43. 236 - 198 38	44. 185 - 173 12	45. 931 - 827 104	46. 679 - 171 508	47. 243 - 225 18	48. 485 - 330 155

Name:................................   Date:................................

Find the difference

Complete all the activities (Subtraction)

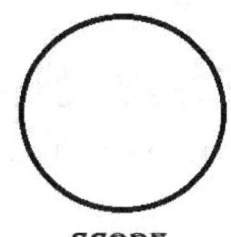

SCORE

1. 738 - 171 --------	2. 680 - 397 --------	3. 141 - 138 --------	4. 390 - 261 --------	5. 840 - 727 --------	6. 629 - 150 --------	7. 423 - 400 --------	8. 115 - 108 --------
9. 161 - 159 --------	10. 739 - 488 --------	11. 251 - 117 --------	12. 613 - 472 --------	13. 515 - 343 --------	14. 315 - 171 --------	15. 561 - 205 --------	16. 295 - 274 --------
17. 406 - 249 --------	18. 733 - 696 --------	19. 830 - 121 --------	20. 211 - 210 --------	21. 762 - 118 --------	22. 843 - 326 --------	23. 215 - 215 --------	24. 545 - 543 --------
25. 872 - 173 --------	26. 150 - 110 --------	27. 244 - 218 --------	28. 485 - 329 --------	29. 586 - 547 --------	30. 145 - 121 --------	31. 727 - 723 --------	32. 139 - 123 --------
33. 666 - 495 --------	34. 779 - 693 --------	35. 301 - 115 --------	36. 763 - 126 --------	37. 852 - 644 --------	38. 883 - 211 --------	39. 359 - 169 --------	40. 636 - 348 --------
41. 181 - 120 --------	42. 523 - 277 --------	43. 256 - 103 --------	44. 756 - 441 --------	45. 153 - 106 --------	46. 459 - 420 --------	47. 980 - 683 --------	48. 361 - 294 --------

Name:................................   Date:................................

Find the difference

Complete all the activities (Subtraction)

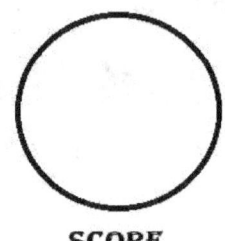

SCORE

1. 738 - 171 567	2. 680 - 397 283	3. 141 - 138 3	4. 390 - 261 129	5. 840 - 727 113	6. 629 - 150 479	7. 423 - 400 23	8. 115 - 108 7
9. 161 - 159 2	10. 739 - 488 251	11. 251 - 117 134	12. 613 - 472 141	13. 515 - 343 172	14. 315 - 171 144	15. 561 - 205 356	16. 295 - 274 21
17. 406 - 249 157	18. 733 - 696 37	19. 830 - 121 709	20. 211 - 210 1	21. 762 - 118 644	22. 843 - 326 517	23. 215 - 215 0	24. 545 - 543 2
25. 872 - 173 699	26. 150 - 110 40	27. 244 - 218 26	28. 485 - 329 156	29. 586 - 547 39	30. 145 - 121 24	31. 727 - 723 4	32. 139 - 123 16
33. 666 - 495 171	34. 779 - 693 86	35. 301 - 115 186	36. 763 - 126 637	37. 852 - 644 208	38. 883 - 211 672	39. 359 - 169 190	40. 636 - 348 288
41. 181 - 120 61	42. 523 - 277 246	43. 256 - 103 153	44. 756 - 441 315	45. 153 - 106 47	46. 459 - 420 39	47. 980 - 683 297	48. 361 - 294 67

Find the difference

Complete all the activities (Subtraction)

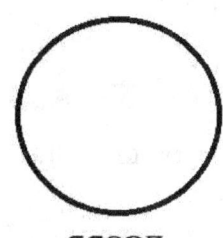
Name:.................................   Date:...................................

1. 983 - 305 --------	2. 707 - 361 --------	3. 319 - 236 --------	4. 794 - 643 --------	5. 420 - 105 --------	6. 426 - 243 --------	7. 954 - 344 --------	8. 377 - 176 --------
9. 391 - 271 --------	10. 434 - 252 --------	11. 570 - 394 --------	12. 479 - 332 --------	13. 971 - 305 --------	14. 832 - 185 --------	15. 708 - 615 --------	16. 154 - 109 --------
17. 489 - 150 --------	18. 243 - 108 --------	19. 632 - 189 --------	20. 297 - 192 --------	21. 523 - 391 --------	22. 156 - 109 --------	23. 689 - 272 --------	24. 612 - 495 --------
25. 820 - 190 --------	26. 610 - 300 --------	27. 362 - 319 --------	28. 297 - 171 --------	29. 828 - 145 --------	30. 514 - 103 --------	31. 826 - 576 --------	32. 196 - 149 --------
33. 959 - 263 --------	34. 104 - 103 --------	35. 214 - 129 --------	36. 714 - 196 --------	37. 430 - 398 --------	38. 984 - 838 --------	39. 606 - 233 --------	40. 961 - 683 --------
41. 699 - 125 --------	42. 384 - 256 --------	43. 351 - 268 --------	44. 893 - 573 --------	45. 917 - 755 --------	46. 321 - 217 --------	47. 735 - 235 --------	48. 694 - 584 --------

Name:................................   Date:................................

Find the difference

Complete all the activities (Subtraction)

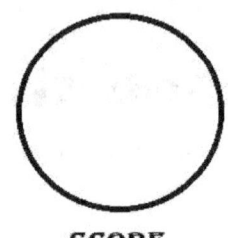

SCORE

1.	983 - 305 678	2.	707 - 361 346	3.	319 - 236 83	4.	794 - 643 151	5.	420 - 105 315	6.	426 - 243 183	7.	954 - 344 610	8.	377 - 176 201

9. 391   10. 434   11. 570   12. 479   13. 971   14. 832   15. 708   16. 154
   - 271   - 252   - 394   - 332   - 305   - 185   - 615   - 109
    120   182   176   147   666   647   93   45

17. 489   18. 243   19. 632   20. 297   21. 523   22. 156   23. 689   24. 612
   - 150   - 108   - 189   - 192   - 391   - 109   - 272   - 495
    339   135   443   105   132   47   417   117

25. 820   26. 610   27. 362   28. 297   29. 828   30. 514   31. 826   32. 196
   - 190   - 300   - 319   - 171   - 145   - 103   - 576   - 149
    630   310   43   126   683   411   250   47

33. 959   34. 104   35. 214   36. 714   37. 430   38. 984   39. 606   40. 961
   - 263   - 103   - 129   - 196   - 398   - 838   - 233   - 683
    696   1   85   518   32   146   373   278

41. 699   42. 384   43. 351   44. 893   45. 917   46. 321   47. 735   48. 694
   - 125   - 256   - 268   - 573   - 755   - 217   - 235   - 584
    574   128   83   320   162   104   500   110

Find the difference

Complete all the activities (Subtraction)

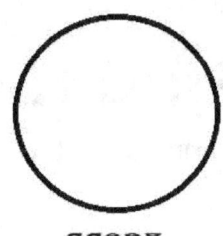

SCORE

1.	622 - 388 --------	2.	282 - 185 --------	3.	359 - 294 --------	4.	681 - 419 --------	5.	497 - 263 --------	6.	123 - 109 --------	7.	867 - 395 --------	8.	815 - 491 --------

9.   194   10.   407   11.   406   12.   308   13.   377   14.   921   15.   105   16.   505
   - 174   - 335   - 270   - 160   - 213   - 367   - 101   - 163
   --------   --------   --------   --------   --------   --------   --------   --------

17.   318   18.   329   19.   155   20.   450   21.   456   22.   199   23.   677   24.   769
   - 137   - 102   - 127   - 421   - 373   - 186   - 385   - 451
   --------   --------   --------   --------   --------   --------   --------   --------

25.   105   26.   482   27.   883   28.   526   29.   389   30.   241   31.   851   32.   588
   - 103   - 453   - 800   - 111   - 364   - 224   - 341   - 240
   --------   --------   --------   --------   --------   --------   --------   --------

33.   964   34.   231   35.   217   36.   536   37.   455   38.   433   39.   501   40.   680
   - 852   - 190   - 192   - 367   - 250   - 291   - 245   - 573
   --------   --------   --------   --------   --------   --------   --------   --------

41.   697   42.   387   43.   756   44.   356   45.   725   46.   719   47.   402   48.   174
   - 332   - 188   - 446   - 189   - 243   - 191   - 254   - 106
   --------   --------   --------   --------   --------   --------   --------   --------

Find the difference

Complete all the activities (Subtraction)

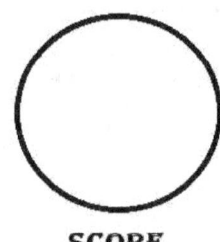

1. 622 - 388 = 234	2. 282 - 185 = 97	3. 359 - 294 = 65	4. 681 - 419 = 262	5. 497 - 263 = 234	6. 123 - 109 = 14	7. 867 - 395 = 472	8. 815 - 491 = 324

1. 622
    - 388

    234

2. 282
    - 185

    97

3. 359
    - 294

    65

4. 681
    - 419

    262

5. 497
    - 263

    234

6. 123
    - 109

    14

7. 867
    - 395

    472

8. 815
    - 491

    324

9. 194
    - 174

    20

10. 407
    - 335

    72

11. 406
    - 270

    136

12. 308
    - 160

    148

13. 377
    - 213

    164

14. 921
    - 367

    554

15. 105
    - 101

    4

16. 505
    - 163

    342

17. 318
    - 137

    181

18. 329
    - 102

    227

19. 155
    - 127

    28

20. 450
    - 421

    29

21. 456
    - 373

    83

22. 199
    - 186

    13

23. 677
    - 385

    292

24. 769
    - 451

    318

25. 105
    - 103

    2

26. 482
    - 453

    29

27. 883
    - 800

    83

28. 526
    - 111

    415

29. 389
    - 364

    25

30. 241
    - 224

    17

31. 851
    - 341

    510

32. 588
    - 240

    348

33. 964
    - 852

    112

34. 231
    - 190

    41

35. 217
    - 192

    25

36. 536
    - 367

    169

37. 455
    - 250

    205

38. 433
    - 291

    142

39. 501
    - 245

    256

40. 680
    - 573

    107

41. 697
    - 332

    365

42. 387
    - 188

    199

43. 756
    - 446

    310

44. 356
    - 189

    167

45. 725
    - 243

    482

46. 719
    - 191

    528

47. 402
    - 254

    148

48. 174
    - 106

    68

Find the difference

Complete all the activities (Subtraction)

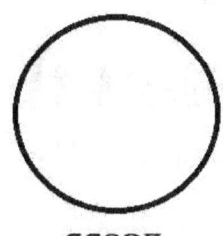

SCORE

1.   551
   - 454
  --------- 	
	2. 614
	   - 370
	  --------- 	
		3. 429
		   - 372
		  --------- 	
			4. 172
			   - 120
			  --------- 	
				5. 977
				   - 359
				  --------- 	
					6. 648
					   - 321
					  --------- 	
						7. 910
						   - 256
						  --------- 	
							8. 366
							   - 125
							  --------- 	

9.   535
   - 475
  --------- 	
	10. 126
	   - 112
	  --------- 	
		11. 121
		   - 116
		  --------- 	
			12. 294
			   - 254
			  --------- 	
				13. 745
				   - 716
				  --------- 	
					14. 252
					   - 201
					  --------- 	
						15. 321
						   - 106
						  --------- 	
							16. 508
							   - 242
							  --------- 	

17.   701
   - 447
  --------- 	
	18. 620
	   - 244
	  --------- 	
		19. 584
		   - 408
		  --------- 	
			20. 948
			   - 826
			  --------- 	
				21. 568
				   - 557
				  --------- 	
					22. 695
					   - 215
					  --------- 	
						23. 580
						   - 161
						  --------- 	
							24. 968
							   - 396
							  --------- 	

25.   809
   - 391
  --------- 	
	26. 718
	   - 523
	  --------- 	
		27. 758
		   - 249
		  --------- 	
			28. 136
			   - 126
			  --------- 	
				29. 612
				   - 500
				  --------- 	
					30. 827
					   - 399
					  --------- 	
						31. 691
						   - 346
						  --------- 	
							32. 129
							   - 127
							  --------- 	

33.   857
   - 297
  --------- 	
	34. 760
	   - 314
	  --------- 	
		35. 651
		   - 111
		  --------- 	
			36. 600
			   - 523
			  --------- 	
				37. 927
				   - 243
				  --------- 	
					38. 929
					   - 139
					  --------- 	
						39. 200
						   - 102
						  --------- 	
							40. 331
							   - 236
							  --------- 	

41.   989
   - 836
  --------- 	
	42. 887
	   - 757
	  --------- 	
		43. 407
		   - 336
		  --------- 	
			44. 654
			   - 345
			  --------- 	
				45. 583
				   - 511
				  --------- 	
					46. 736
					   - 582
					  --------- 	
						47. 845
						   - 496
						  --------- 	
							48. 278
							   - 119
							  --------- 	

Name:.............................. Date:..............................

Find the difference

Complete all the activities (Subtraction)

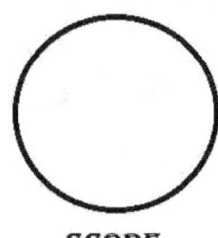

SCORE

| 1. | 551
- 454
97 | 2. | 614
- 370
244 | 3. | 429
- 372
57 | 4. | 172
- 120
52 | 5. | 977
- 359
618 | 6. | 648
- 321
327 | 7. | 910
- 256
654 | 8. | 366
- 125
241 |

9.   535   10. 126   11. 121   12. 294   13. 745   14. 252   15. 321   16. 508
  - 475   - 112   - 116   - 254   - 716   - 201   - 106   - 242
    60   14   5   40   29   51   215   266

17. 701   18. 620   19. 584   20. 948   21. 568   22. 695   23. 580   24. 968
  - 447   - 244   - 408   - 826   - 557   - 215   - 161   - 396
    254   376   176   122   11   480   419   572

25. 809   26. 718   27. 758   28. 136   29. 612   30. 827   31. 691   32. 129
  - 391   - 523   - 249   - 126   - 500   - 399   - 346   - 127
    418   195   509   10   112   428   345   2

33. 857   34. 760   35. 651   36. 600   37. 927   38. 929   39. 200   40. 331
  - 297   - 314   - 111   - 523   - 243   - 139   - 102   - 236
    560   446   540   77   684   790   98   95

41. 989   42. 887   43. 407   44. 654   45. 583   46. 736   47. 845   48. 278
  - 836   - 757   - 336   - 345   - 511   - 582   - 496   - 119
    153   130   71   309   72   154   349   159

Name:.................................   Date:.................................

Find the difference

Complete all the activities (Subtraction)

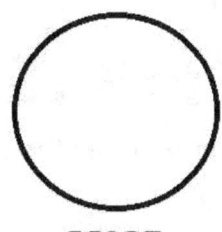

SCORE

1. 950 - 708 -------	2. 415 - 186 -------	3. 454 - 327 -------	4. 754 - 603 -------	5. 868 - 746 -------	6. 787 - 484 -------	7. 373 - 371 -------	8. 655 - 403 -------
9. 570 - 203 -------	10. 480 - 415 -------	11. 586 - 196 -------	12. 644 - 140 -------	13. 679 - 269 -------	14. 807 - 640 -------	15. 381 - 324 -------	16. 116 - 106 -------
17. 976 - 816 -------	18. 191 - 136 -------	19. 705 - 695 -------	20. 276 - 161 -------	21. 190 - 110 -------	22. 981 - 541 -------	23. 105 - 100 -------	24. 224 - 167 -------
25. 310 - 134 -------	26. 311 - 229 -------	27. 198 - 136 -------	28. 606 - 479 -------	29. 199 - 128 -------	30. 666 - 157 -------	31. 869 - 842 -------	32. 511 - 123 -------
33. 840 - 379 -------	34. 626 - 356 -------	35. 903 - 277 -------	36. 886 - 524 -------	37. 121 - 102 -------	38. 756 - 385 -------	39. 295 - 120 -------	40. 160 - 135 -------
41. 239 - 134 -------	42. 577 - 172 -------	43. 607 - 307 -------	44. 350 - 251 -------	45. 329 - 181 -------	46. 357 - 201 -------	47. 418 - 130 -------	48. 541 - 160 -------

© KingSchool Edition

Find the difference

Complete all the activities (Subtraction)

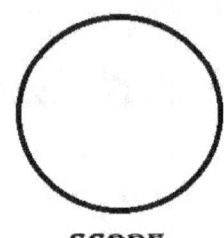

SCORE

1.	950	2.	415	3.	454	4.	754	5.	868	6.	787	7.	373	8.	655
	- 708		- 186		- 327		- 603		- 746		- 484		- 371		- 403
	242		229		127		151		122		303		2		252

9.	570	10.	480	11.	586	12.	644	13.	679	14.	807	15.	381	16.	116
	- 203		- 415		- 196		- 140		- 269		- 640		- 324		- 106
	367		65		390		504		410		167		57		10

17.	976	18.	191	19.	705	20.	276	21.	190	22.	981	23.	105	24.	224
	- 816		- 136		- 695		- 161		- 110		- 541		- 100		- 167
	160		55		10		115		80		440		5		57

25.	310	26.	311	27.	198	28.	606	29.	199	30.	666	31.	869	32.	511
	- 134		- 229		- 136		- 479		- 128		- 157		- 842		- 123
	176		82		62		127		71		509		27		388

33.	840	34.	626	35.	903	36.	886	37.	121	38.	756	39.	295	40.	160
	- 379		- 356		- 277		- 524		- 102		- 385		- 120		- 135
	461		270		626		362		19		371		175		25

41.	239	42.	577	43.	607	44.	350	45.	329	46.	357	47.	418	48.	541
	- 134		- 172		- 307		- 251		- 181		- 201		- 130		- 160
	105		405		300		99		148		156		288		381

Name:................................   Date:.................................

Find the difference

Complete all the activities (Subtraction)

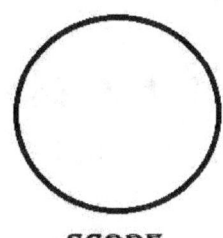

SCORE

| 1. | 683
- 639
-------- | 2. | 709
- 483
-------- | 3. | 524
- 377
-------- | 4. | 684
- 107
-------- | 5. | 358
- 279
-------- | 6. | 627
- 117
-------- | 7. | 675
- 174
-------- | 8. | 579
- 103
-------- |

| 9. | 289
- 108
-------- | 10. | 951
- 620
-------- | 11. | 490
- 219
-------- | 12. | 543
- 284
-------- | 13. | 188
- 186
-------- | 14. | 308
- 212
-------- | 15. | 941
- 503
-------- | 16. | 268
- 256
-------- |

| 17. | 627
- 413
-------- | 18. | 807
- 307
-------- | 19. | 484
- 268
-------- | 20. | 204
- 152
-------- | 21. | 634
- 376
-------- | 22. | 878
- 776
-------- | 23. | 839
- 281
-------- | 24. | 968
- 210
-------- |

| 25. | 837
- 306
-------- | 26. | 265
- 151
-------- | 27. | 234
- 218
-------- | 28. | 731
- 316
-------- | 29. | 487
- 202
-------- | 30. | 577
- 334
-------- | 31. | 478
- 141
-------- | 32. | 170
- 157
-------- |

| 33. | 557
- 401
-------- | 34. | 502
- 119
-------- | 35. | 407
- 174
-------- | 36. | 620
- 502
-------- | 37. | 934
- 471
-------- | 38. | 116
- 115
-------- | 39. | 496
- 347
-------- | 40. | 633
- 554
-------- |

| 41. | 113
- 101
-------- | 42. | 748
- 524
-------- | 43. | 366
- 333
-------- | 44. | 801
- 595
-------- | 45. | 901
- 455
-------- | 46. | 874
- 228
-------- | 47. | 916
- 609
-------- | 48. | 173
- 135
-------- |

Name:................................ Date:................................

Find the difference

Complete all the activities (Subtraction)

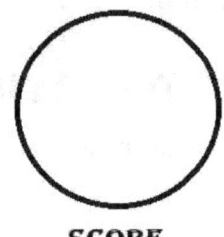

SCORE

1. 683 - 639 44	2. 709 - 483 226	3. 524 - 377 147	4. 684 - 107 577	5. 358 - 279 79	6. 627 - 117 510	7. 675 - 174 501	8. 579 - 103 476
9. 289 - 108 181	10. 951 - 620 331	11. 490 - 219 271	12. 543 - 284 259	13. 188 - 186 2	14. 308 - 212 96	15. 941 - 503 438	16. 268 - 256 12
17. 627 - 413 214	18. 807 - 307 500	19. 484 - 268 216	20. 204 - 152 52	21. 634 - 376 258	22. 878 - 776 102	23. 839 - 281 558	24. 968 - 210 758
25. 837 - 306 531	26. 265 - 151 114	27. 234 - 218 16	28. 731 - 316 415	29. 487 - 202 285	30. 577 - 334 243	31. 478 - 141 337	32. 170 - 157 13
33. 557 - 401 156	34. 502 - 119 383	35. 407 - 174 233	36. 620 - 502 118	37. 934 - 471 463	38. 116 - 115 1	39. 496 - 347 149	40. 633 - 554 79
41. 113 - 101 12	42. 748 - 524 224	43. 366 - 333 33	44. 801 - 595 206	45. 901 - 455 446	46. 874 - 228 646	47. 916 - 609 307	48. 173 - 135 38

Find the difference

Complete all the activities (Subtraction)

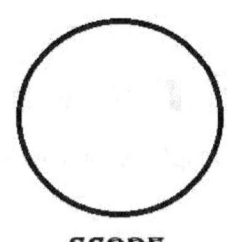

1. 496 - 123 --------	2. 425 - 153 --------	3. 818 - 760 --------	4. 805 - 509 --------	5. 731 - 360 --------	6. 443 - 342 --------	7. 875 - 709 --------	8. 911 - 755 --------
9. 702 - 508 --------	10. 529 - 235 --------	11. 449 - 247 --------	12. 560 - 390 --------	13. 890 - 685 --------	14. 239 - 225 --------	15. 440 - 240 --------	16. 877 - 336 --------
17. 427 - 274 --------	18. 528 - 264 --------	19. 336 - 201 --------	20. 149 - 105 --------	21. 388 - 140 --------	22. 956 - 857 --------	23. 545 - 133 --------	24. 859 - 180 --------
25. 587 - 126 --------	26. 545 - 202 --------	27. 232 - 201 --------	28. 187 - 102 --------	29. 591 - 458 --------	30. 356 - 115 --------	31. 787 - 709 --------	32. 282 - 153 --------
33. 592 - 138 --------	34. 406 - 201 --------	35. 378 - 244 --------	36. 590 - 218 --------	37. 661 - 315 --------	38. 994 - 322 --------	39. 523 - 365 --------	40. 223 - 199 --------
41. 906 - 140 --------	42. 421 - 204 --------	43. 925 - 624 --------	44. 242 - 185 --------	45. 492 - 280 --------	46. 853 - 501 --------	47. 230 - 230 --------	48. 397 - 229 --------

Find the difference

Complete all the activities (Subtraction)

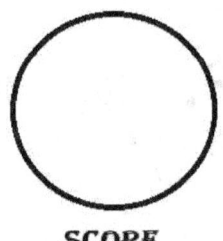

SCORE

1.	496	2.	425	3.	818	4.	805	5.	731	6.	443	7.	875	8.	911
	- 123		- 153		- 760		- 509		- 360		- 342		- 709		- 755
	373		272		58		296		371		101		166		156

9.	702	10.	529	11.	449	12.	560	13.	890	14.	239	15.	440	16.	877
	- 508		- 235		- 247		- 390		- 685		- 225		- 240		- 336
	194		294		202		170		205		14		200		541

17.	427	18.	528	19.	336	20.	149	21.	388	22.	956	23.	545	24.	859
	- 274		- 264		- 201		- 105		- 140		- 857		- 133		- 180
	153		264		135		44		248		99		412		679

25.	587	26.	545	27.	232	28.	187	29.	591	30.	356	31.	787	32.	282
	- 126		- 202		- 201		- 102		- 458		- 115		- 709		- 153
	461		343		31		85		133		241		78		129

33.	592	34.	406	35.	378	36.	590	37.	661	38.	994	39.	523	40.	223
	- 138		- 201		- 244		- 218		- 315		- 322		- 365		- 199
	454		205		134		372		346		672		158		24

41.	906	42.	421	43.	925	44.	242	45.	492	46.	853	47.	230	48.	397
	- 140		- 204		- 624		- 185		- 280		- 501		- 230		- 229
	766		217		301		57		212		352		0		168

Find the difference

Complete all the activities (Subtraction)

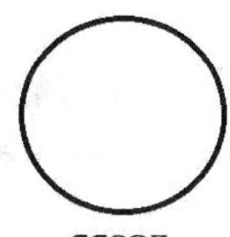

SCORE

Name:............................... Date:...............................

1.	944	2.	403	3.	760	4.	816	5.	557	6.	645	7.	793	8.	592
	- 660		- 226		- 440		- 116		- 216		- 516		- 688		- 552

9.	633	10.	245	11.	576	12.	434	13.	927	14.	903	15.	713	16.	187
	- 356		- 204		- 518		- 405		- 715		- 783		- 108		- 155

17.	147	18.	378	19.	361	20.	162	21.	327	22.	833	23.	126	24.	843
	- 126		- 324		- 288		- 124		- 176		- 587		- 110		- 537

25.	919	26.	850	27.	913	28.	363	29.	546	30.	660	31.	406	32.	241
	- 853		- 561		- 149		- 212		- 542		- 630		- 252		- 238

33.	739	34.	584	35.	664	36.	470	37.	516	38.	156	39.	894	40.	924
	- 234		- 459		- 592		- 389		- 181		- 140		- 518		- 351

41.	945	42.	167	43.	813	44.	690	45.	563	46.	302	47.	781	48.	994
	- 744		- 129		- 412		- 247		- 500		- 116		- 309		- 386

Name:....................................... Date:...............................

Find the difference

Complete all the activities (Subtraction)

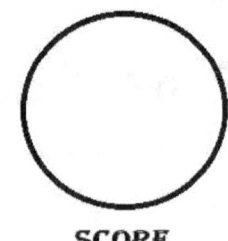

SCORE

1. 944 - 660 = 284	2. 403 - 226 = 177	3. 760 - 440 = 320	4. 816 - 116 = 700	5. 557 - 216 = 341	6. 645 - 516 = 129	7. 793 - 688 = 105	8. 592 - 552 = 40
9. 633 - 356 = 277	10. 245 - 204 = 41	11. 576 - 518 = 58	12. 434 - 405 = 29	13. 927 - 715 = 212	14. 903 - 783 = 120	15. 713 - 108 = 605	16. 187 - 155 = 32
17. 147 - 126 = 21	18. 378 - 324 = 54	19. 361 - 288 = 73	20. 162 - 124 = 38	21. 327 - 176 = 151	22. 833 - 587 = 246	23. 126 - 110 = 16	24. 843 - 537 = 306
25. 919 - 853 = 66	26. 850 - 561 = 289	27. 913 - 149 = 764	28. 363 - 212 = 151	29. 546 - 542 = 4	30. 660 - 630 = 30	31. 406 - 252 = 154	32. 241 - 238 = 3
33. 739 - 234 = 505	34. 584 - 459 = 125	35. 664 - 592 = 72	36. 470 - 389 = 81	37. 516 - 181 = 335	38. 156 - 140 = 16	39. 894 - 518 = 376	40. 924 - 351 = 573
41. 945 - 744 = 201	42. 167 - 129 = 38	43. 813 - 412 = 401	44. 690 - 247 = 443	45. 563 - 500 = 63	46. 302 - 116 = 186	47. 781 - 309 = 472	48. 994 - 386 = 608

www.ingramcontent.com/pod-product-compliance
Lightning Source LLC
Chambersburg PA
CBHW081523220526
45467CB00010B/3022